DATE DUE FOR RETURN

UNIVERSITY LIBRARY
25 JAN 2005
SEM GGL 07

This book may be recalled before the above date.

POPULATION CYCLES

POPULATION CYCLES
The Case for Trophic Interactions

Edited by
ALAN BERRYMAN

UNIVERSITY PRESS

2002

OXFORD
UNIVERSITY PRESS

Oxford New York
Auckland Bangkok Buenos Aires Cape Town Chennai
Dar es Salaam Delhi Hong Kong Istanbul Karachi Kolkata
Kuala Lumpur Madrid Melbourne Mexico City Mumbai Nairobi
São Paulo Shanghai Singapore Taipei Tokyo Toronto

Copyright © 2002 by Oxford University Press, Inc.

Published by Oxford University Press, Inc.
198 Madison Avenue, New York, New York 10016

www.oup.com

Oxford is a registered trademark of Oxford University Press

All rights reserved. No part of this publication may be reproduced,
stored in a retrieval system, or transmitted, in any form or by any means,
electronic, mechanical, photocopying, recording, or otherwise,
without the prior permission of Oxford University Press.

Library of Congress Cataloging-in-Publication Data
Population cycles: the case for trophic interactions / edited by Alan Berryman.
    p. cm.
  Includes bibliographical references.
  ISBN 0-19-514098-2
  1. Animal populations.   I. Berryman, A. A. (Alan Andrew), 1937–
QL752 .P635 2002
591.7'88—dc21   2002070032

9 8 7 6 5 4 3 2 1
Printed in the United States of America
on acid-free paper

In Memory of Charles Elton,

Father of the Population Cycle

Preface

In 1999 I organized a symposium at the annual meeting of the Ecological Society of America in Spokane, Washington. The intent of this symposium was to highlight recent research suggesting that population cycles in *some* species are likely to be caused by trophic interactions or, in a broader sense, by the interaction structure, or architecture, of the ecological system. Of course, this is not a new idea for its origins can be traced back at least as far as the Lotka–Volterra predator–prey models. However, it is an idea that had fallen out of fashion during the middle of this century, as ecologists searched for general causes in intrinsic, self-regulatory processes. The symposium was very well attended and, at times, overflowing. It generated vigorous and sometimes heated debate. Many attendees mentioned that the subject and material were extremely interesting and topical, and would make a good book.

When I agreed to edit this book I thought it would be a fairly simple operation. After all, we had just finished presenting our papers at a symposium, so all I needed to do was gather the papers together and send them off to the publisher. How wrong I was. I have now read each chapter several times and have fretted, perhaps unduly, over them. Some authors, I'm sure, have been frustrated and irritated by my contentious attention to detail and the delays in publication (not entirely my fault, I might add).

The main contention of this book, and many of its authors, is that the numerical dynamics of ecological systems are, in large part, the result of its structure. I suspect that many ecologists would agree with this statement in general, but disagree with it when talking about population cycles. This is rather surprising. What is the mystery about population cycles? Why are they thought to be something special, requiring a completely different explanation from other kinds of population dynamics? Our main argument is that cycles are, in at least *some* animal populations, a consequence of the kind of ecological structure that they find themselves embedded within, and that trophic interactions can create the kind of structure required to generate cyclical dynamics.

The authors of this book were selected because their research in some way or another supports the idea that population cycles can be generated by trophic interactions. Our purpose is not to compare or contrast competing hypotheses, for much has been written on this subject over the last half-century. Rather, we want to present recent evidence for the role of trophic interactions in population cycles. Lest we be misunderstood, let me repeat that we are not proposing a universal explanation (as many have attempted to do), but merely pointing out that there is good evidence to support the idea that trophic interactions are involved in *some* species. The evidence seems pretty convincing in some of the examples discussed in these pages. In others it may be less so. The reader will have to decide just how convincing. To add perspective we have invited criticism by others who may disagree with our methods or conclusions.

Pullman, Washington A. A. B
February, 2002

Contents

Contributors xi

1 Population Cycles: Causes and Analysis 3
   Alan A. Berryman

2 The Role of Insect Parasitoids in Population Cycles of the Spruce Needleminer in Denmark 29
   Mikael Münster-Swendsen

3 Population Cycles of Small Rodents in Fennoscandia 44
   Ilkka Hanski and Heikki Henttonen

4 Understanding the Snowshoe Hare Cycle through Large-scale Field Experiments 69
   Stan Boutin, Charles J. Krebs, Rudy Boonstra, Anthony R. E. Sinclair, and Karen E. Hodges

5 Evidence for Predator–Prey Cycles in a Bark Beetle 92
   John D. Reeve and Peter Turchin

6 Parasitic Worms and Population Cycles of Red Grouse 109
   Peter J. Hudson, Andrew P. Dobson, and David Newborn

7 Population Cycles of the Larch Budmoth in Switzerland   130

Peter Turchin, Cheryl J. Briggs, Stephen P. Ellner, Andreas Fischlin, Bruce E. Kendall, Edward McCauley, William W. Murdoch, and Simon N. Wood

8 Population Cycles of the Autumnal Moth in Fennoscandia   142

Miia Tanhuanpää, Kai Ruohomäki, Peter Turchin, Matthew P. Ayres, Helena Bylund, Pekka Kaitaniemi, Toomas Tammaru, and Erkki Haukioja

9 Population Cycles: Inferences from Experimental, Modeling, and Time Series Approaches   155

Xavier Lambin, Charles J. Krebs, Robert Moss, and Nigel G. Yoccoz

10 Do Trophic Interactions Cause Population Cycles?   177

Alan A. Berryman

Index   189

Contributors

Matthew P. Ayres
  Department of Biological Sciences
  Dartmouth College
  Hanover, New Hampshire 03755-3576
  USA
  matthew.p.ayres@dartmouth.edu

Alan A. Berryman
  Department of Entomology
  Washington State University
  Pullman, Washington 99164
  USA
  berryman@mail.wsu.edu

Rudy Boonstra
  Division of Life Sciences
  University of Toronto at Scarborough
  Scarborough, Ontario M1C 1A4
  Canada
  boonstra@scar.utoronto.ca

Stan Boutin
  Department of Biological Sciences
  University of Alberta
  Edmonton, Alberta T6G 2E9
  Canada
  stan.boutin@ualberta.ca

Cheryl J. Briggs
  Department of Integrative Biology
  University of California
  Berkeley, California 94720
  USA
  cbriggs@socrates.berkeley.edu

Helena Bylund
  Department of Entomology
  Swedish University of Agricultural
    Sciences
  P.O. Box 7044, SE-750 07 Uppsala
  Sweden
  helena.bylund@entom.slu.se

Andrew P. Dobson
  Ecology and Evolutionary Biology
  Princeton University
  Princeton, New Jersey 08540-1003
  USA
  andy@eno.princeton.edu

Stephen P. Ellner
  Department of Ecology and
    Evolutionary Biology
  Corson Hall, Cornell University
  Ithaca, New York 14853
  spe2@cornell.edu

Contributors

Andreas Fischlin
  Institute of Terrestrial Ecology
  Swiss Federal Institute of Technology
  Zurich ETHZ
  CH-8952 Schlieren/Zurich
  Switzerland
  andrea_fischlin@yahoo.com

Ilkka Hanski
  Department of Ecology and
    Systematics
  P.O. Box 65
  FIN-00014 University of Helsinki
  Finland
  ilkka.hanski@helsinki.fi

Erkki Haukioja
  Section of Ecology
  Department of Biology
  University of Turku
  FIN-20014 Turku
  Finland
  erkki.haukioja@utu.fi

Heikki Henttonen
  Vantaa Research Center
  P.O. Box 18
  Forest Research Institute
  FIN-01301 Vantaa
  Finland
  heikki.henttonen@metla.fi

Karen E. Hodges
  Department of Zoology
  University of Aberdeen
  Tillydrone Avenue
  Aberdeen AB24 2TZ
  UK
  khodges@forestry.umt.edu

Peter J. Hudson
  Institute of Biological Sciences
  University of Stirling
  Scotland FK9 4LA
  UK
  p.j.hudson@stir.ac.uk

Pekka Kaitaniemi
  Section of Ecology
  Department of Biology
  University of Turku
  FIN-20014 Turku
  Finland
  pekka.kaitaniemi@utu.fi

Bruce E. Kendall
  School of Environmental Science and
    Management
  University of California
  Santa Barbara, California 93106
  USA
  kendall@bren.ucsb.edu

Charles J. Krebs
  Department of Zoology
  University of British Colombia
  6270 University Boulevard
  Vancouver, British Columbia V6T 1Z4
  Canada
  krebs@zoology.ubc.ca

Xavier Lambin
  Department of Zoology
  University of Aberdeen
  Tillydrone Avenue
  Aberdeen AB24 2TZ
  UK
  x.lambin@abdn.ac.uk

Edward McCauley
  Ecology Division
  Department of Biological Sciences
  University of Calgary
  Calgary T2N 1N4
  Canada
  mccauley@acs.ucalgary.ca

Robert Moss
  Centre for Ecology and Hydrology
  Hill of Brathens
  Glassel, Banchory
  Kincardineshire AB31 4BY
  UK
  rmoss@ceh.ac.uk

Mikael Münster-Swendsen
  Department of Population Ecology
  Zoological Institute
  University of Copenhagen
  Universitetsparken 15

DK-2100 Copenhagen
Denmark
mmswendsen@zi.ku.dk

William W. Murdoch
Department of Ecology, Evolution,
and Marine Biology
University of California
Santa Barbara, California 93106
USA
murdoch@lifesci.lscf.ucsb.edu

David Newborn
The Game Conservancy Trust
Swale Farm
Satron, Gunnerside
Richmond, North Yorks DL11 3JW
UK
dnewborn@gct.org.uk

John D. Reeve
Department of Zoology
Southern Illinois University
Carbondale, Illinois 62901-6501
USA
jreeve@zoology.siu.edu

Kai Ruohomäki
Section of Ecology
Department of Biology
University of Turku
FIN-20014 Turku
Finland
kai.ruohomaki@utu.fi

Anthony R. E. Sinclair
Centre for Biodiversity Research
University of British Columbia
6270 University Boulevard
Vancouver, British Columbia V6T 1Z4
Canada
sinclair@zoology.ubc.ca

Toomas Tammaru
Institute of Zoology and Botany
Riia 181
EE-51014 Tartu
Estonia
toomast@zbi.ee

Miia Tanhuanpää
Section of Ecology
Department of Biology
University of Turku
FIN-20014 Turku
Finland
miia.tanhuanpaa@utu.fi

Peter Turchin
Department of Ecology and
Evolutionary Biology
University of Connecticut
Storrs, Connecticut 06269-3042
USA
peter.turchin@uconn.edu

Simon N. Wood
Mathematical Institute
North Haugh
St. Andrews, Fife
KY16 9SS
UK
snw@st.andrews.ac.uk

Nigel G. Yoccoz
Norwegian Institute for Nature
Research
Polar Environmental Centre
9005 Tromsø
Norway
nigel.yoccoz@ninatos.ninaniku.no

POPULATION CYCLES

1

Population Cycles
Causes and Analysis

Alan A. Berryman

1.1 Introduction

Ever since Elton's classic book *Voles, Mice and Lemmings* (Elton 1942), understanding and explaining the causes of regular multiannual cycles in animal populations has been a central issue in ecology. Many hypotheses have been erected and incessantly argued about, but no clear picture has emerged. Below I briefly sketch the major hypotheses without any attempt to be complete or to comment on their relative merits or demerits. Detailed reviews and discussion can be found in Keith (1963), Krebs and Myers (1974), Finerty (1980), Myers (1988), Royama (1992), and Stenseth (1999).

(H1) *Physical effects* (e.g., Elton 1924, Bodenheimer 1938). Perhaps the most obvious hypothesis is that cycles in animal populations reflect the response of birth and death rates to an external physical factor that is itself cyclic. Two of the more specific physical hypotheses involve periodic climatic factors and sunspot activity.

(H2) *Predator effects*. Lotka (1924) and Volterra (1926) demonstrated that cyclic dynamics are inherent in simple predator–prey models, leading to the hypothesis that regular cycles can result from interactions between predator and prey populations.

(H3) *Pathogen effects*. Anderson and May (1980) showed that, under certain conditions, simple models of infectious disease transmission can generate cycles in host and pathogen populations. This is similar to H2 with the pathogen as a predator.

(H4) *Plant effects*. Several hypotheses have been proposed for the possible role of plants in generating population cycles of herbivores. One is a generalization of H2 in which the plant is considered the prey and the herbivore the predator (Elton 1924, Pitelka 1957). Another involves nutrient cycling: In this hypothesis, nutrient deficiencies are assumed to reduce the resistance of plants, resulting in larger herbivore populations, but nutrients released in feces and decaying animal and plant matter cycle back to the plants, increasing their vigor and resistance, and resulting in reduced herbivory (e.g., White 1974). Another hypothesis argues that herbivore feeding induces sustained chemical and/or physical changes in the plant (delayed induced resistance), which then reduce the reproduction and/or survival of future herbivore generations (Benz 1974, Haukioja and Hakala 1975).

(H5) *Maternal effects*. This hypothesis proposes that cycles can be caused by qualitative changes in individuals due to stress on the maternal generation (Christian 1950, Wellington 1960). For example, the offspring of well-fed mothers could be more vigorous and have higher survival and reproductive rates than those of undernourished mothers (see also Rossiter 1991, Ginzburg and Taneyhill 1994). This hypothesis may be related to certain aspects of the "plant effect" hypothesis (H4 above) because of the causal link between food quantity/quality and maternal nutrition, and to H2 and H3 because of the causal link between weak offspring and susceptibility to predators and/or pathogens.

(H6) *Genetic effects*. Chitty (1957, 1967) proposed that cyclic population dynamics could result from density-induced changes in genetic traits affecting survival and/or reproduction of individual organisms. According to this argument, natural selection favors individuals with lower fecundity and/or survival when populations are dense, and higher fecundity and/or survival when populations are sparse, and this density-dependent natural selection causes populations to cycle (Witting 1997). A variant is the kin-selection hypothesis, where genetically related individuals are more tolerant of one another so that higher population densities are possible (Moss and Watson 1991).

In the beginning, the search for causes of population cycles centered on physical factors and the effects of predators (H1 and H2). Then, with the publication of Chitty's ideas in the 1960s, attention switched to the genetic hypothesis (H6) (for a personal account of this interesting debate, see Chitty 1996). In the early 1980s, the theoretical models of Anderson and May focused attention on pathogens as a possible cause of cyclic dynamics in their prey populations (H3), and research on plant–herbivore interactions generated considerable interest in plant effect (H4) and maternal effect (H5) hypotheses. Finally, the last decade has seen a refocusing on the effects of food, predators, and pathogens (H2, H3, and H4) or, more generally, on the role of trophic interactions or food web architecture in population cycles. The major purpose of this book is to review these latter developments. I should make it clear, however, that we are not proposing a *general* explanation for all population cycles, but merely trying to present evidence that population cycles are likely to be caused by food web architecture in *some* natural systems. Claims for a general (universal) biological explanation of population cycles are, in my opinion, premature. As we will see below, there are many specific biological mechanisms that can induce cyclic dynamics in animal populations, and these can change from time to time and from place to place.

1.2 What Is a Population Cycle?

At this point it is important to clearly define what we mean in this book by a "population cycle," as there is some confusion on this issue (e.g., see Hunter and Price 1998, Turchin and Berryman 2000). In a strictly mathematical and statistical sense, a cycle can be defined as any oscillation that has a statistically significant periodicity (Finerty 1980). From this point of view, both the regular 2-year oscillations observed in populations of sycamore aphids and the 9-year oscillation of larch budmoths (figure 1.1) can be considered cyclical because both exhibit statistically significant periodicity (see section 1.4 for statistical procedures). There are, however, a couple of problems with this strict mathematical definition.

The first problem is semantic. While mathematicians define any periodic oscillation as a cycle, ecologists usually reserve the term for an oscillation that takes several or many years (i.e., 3 years or more) to repeat itself. Hence, most ecologists would consider the larch budmoth population to be cyclic and the sycamore aphid population to be noncyclic. Although this distinction may not make sense to a mathematician, it does to an ecologist because different ecological processes are usually involved. For example, 2-year oscillations are typical of populations regulated by intrapopulation processes (e.g., competition for food and space), while multiannual cycles are often seen in populations regulated by interpopulation processes (e.g., interactions with food or predators; see section 1.3).

The second problem is statistical. Ecological time series are often short and variable, which sometimes makes it difficult to demonstrate statistical significance. For example, the spruce needleminer time series (chapter 2) does not have a statistically significant period, even though the 19-year time series, together with qualitative historical records, clearly shows that population peaks are spatially synchronized and occur at fairly regular intervals (see figures 2.1 and 2.2). Thus, because this book is about ecology and for ecologists, we will define a population cycle as an oscillation in population numbers or density that has an obviously regular period of three or more years.

1.3 What Causes Population Cycles?

The problem of causation can be approached by recognizing two levels of explanation: General explanations that apply to all dynamic systems (i.e., they are not system-specific) and specific explanations that apply to a particular system (e.g., the larch budmoth).

6 Population Cycles

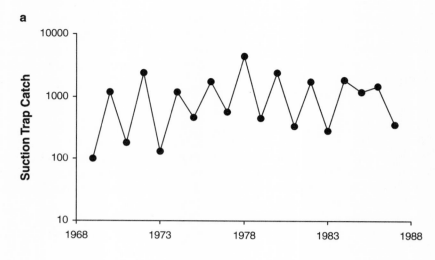

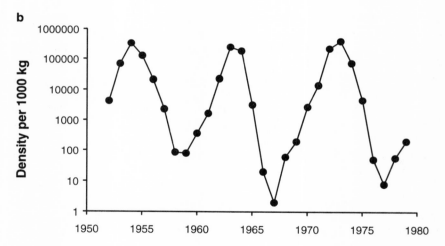

Figure 1.1 Numbers of (a) sycamore aphids caught annually in a suction trap (data from Dixon 1990) and (b) larch budmoth larvae counted annually per 1000 kg of larch foliage (data from Baltensweiler 1989).

The most obvious general cause of cycles in any system is that they follow, or are driven by, an external variable that is itself cyclic. We call these exogenous cycles.

A second general cause involves the idea of delayed negative feedback (Hutchinson 1948, Morris 1959, Varley et al. 1973, Berryman 1981, 1989). One of the central tenets of control theory is that time delays in negative feedback loops can cause oscillatory instability and regular cycles in the variables involved in the loop (Milsum 1966). In addition, it is generally understood that the length of the time delay is directly related to the number of variables in the loop. Thus, cycles can be caused by negative feedback loops that involve two or more dynamic variables. We call these endogenous cycles because they result from the internal structural properties (architecture) of the system.

1.3.1 Specific Exogenous Mechanisms

As noted above, the term *exogenous cycle* is used to describe a multiannual periodic oscillation caused by the action of an independent external driving variable(s). Thus, the dynamics observed in a particular variable(s) are the result of a unidirectional causal process (figure 1.2)

$$E \to A \to B \to \cdots \to C, \tag{1.1}$$

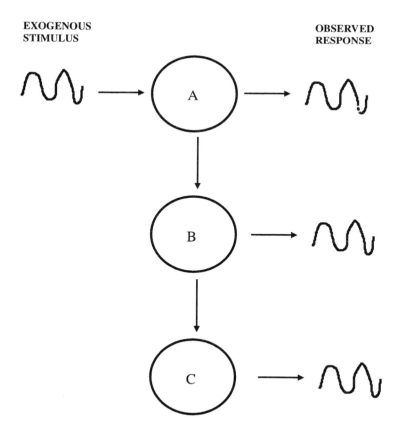

Figure 1.2 Linear causal system in which the observed cycles in variables A, B, and C are caused by a cyclic exogenous factor.

where E is a cyclic exogenous variable, or group of coacting interdependent exogenous variables, and A, B, and C are response variables, one or more of which is measured or observed. Notice that cyclic dynamics in any one variable (say C) may be the result of exogenous forces acting on another variable (say A) to which it is causally linked. To obtain a plausible explanation for an exogenous population cycle, one first needs to identify a logical chain of causation, and then to demonstrate that the causal exogenous variable(s) dominates the dynamics of the response variable(s). An example of a cycle caused by an exogenous variable comes from my work with the fir engraver beetle (Berryman 1973). Populations of this bark beetle fluctuate in response to the abundance of physiologically stressed fir trees, which in turn fluctuates periodically due to the impact of a cyclic defoliator, the Douglas-fir tussock moth (Wright et al. 1984). Hence, fir engraver population cycles (say variable C in figure 1.2) are driven by oscillations in the abundance of susceptible hosts (B), which are caused by cyclic fluctuations in Douglas-fir tussock moth populations (A). The causes of population cycles in the tussock moth are another story.

1.3.2 Specific Endogenous Mechanisms

The term *endogenous cycle* is used to describe a multiannual periodic oscillation that results from the action of a delayed negative feedback loop, as defined in general by the circular causal process (figure 1.3)

$$A \rightarrow B \rightarrow \cdots \rightarrow A. \tag{1.2}$$

Here, there is no need to specify an exogenous effect because the cycle is a result of the structure of the endogenous system. The number of dynamic variables in the loop defines the *order* of the endogenous process; for example, the feedback between A and B in figure 1.3 is of second order because it involves two dynamic variables. *The necessary (but not sufficient) conditions for an endogenous multiannual cycle are that the feedback loop be at least second order and the sign of the loop be negative.* The latter condition is fulfilled if the product of all the interactions in the loop is negative. For example, the second-order feedback loop in figure 1.3 is negative if, and only if, one of the interaction effects (arrows) has a negative sign and the other has a positive sign.

Although dynamic systems fulfilling these necessary conditions have the potential to exhibit cycles, the actual observation of regular oscillations may require other (sufficient) conditions to be met. For example, a system governed by second-order negative feedback may damp down to a stable equilibrium in a constant environment, but cycle in a variable environment (Berryman 1986, 1999b; Bjørnstad et al. 1995). However, as all biological systems are affected by environmental variability, we can usually assume that this condition will be met.

Specific mechanisms for endogenous cycles must explain how the dynamic variables interact to create the necessary delayed negative feedback architec-

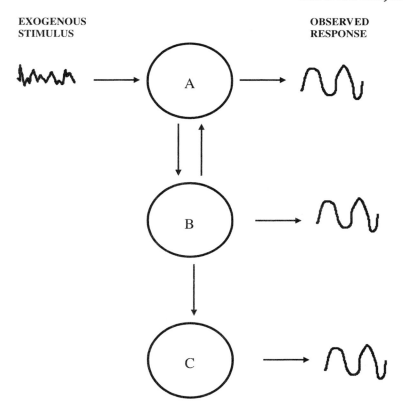

Figure 1.3 Circular causal system in which the observed cycles in variables A, B, and C are caused by an endogenous interaction between variables A and B.

ture. In the case of biological populations, these mechanisms (H2–H6) fall into two main groups.

(1) *Intrapopulation (self-regulating) mechanisms*. Delayed negative feedback can be created if crowded conditions within a population result in lower survival and/or reproduction of offspring in succeeding generations. The observed response variable (say A in figure 1.3) is usually the density of the population of interest, and the causal variable B was often an average property of the individuals within the population. For example, the variable B could be the frequency of sluggish, aggressive, dark-colored, or related individuals (Wellington 1960, Chitty 1967, Baltensweiler 1978, Moss and Watson 1991). If B is an inherited genetic trait, then a specific explanation would involve the "genetic effects" hypothesis (H6), while if it was a phenotypic trait, it would involve the "maternal effects" hypothesis (H5). Both genetic and maternal effects can create time delays in the feedback structure because the reproduction and survival of one generation is affected by the density of a previous generation(s). In order to propose a plausible genetic or maternal

effects hypothesis, however, one must demonstrate a strong positive (or direct) effect of the observed variable (A) on the causal variable (B) and a strong negative (or inverse) effect of B on A, or vice versa. It is important to realize that the identification of such a negative feedback loop is not, in itself, sufficient proof of the self-regulation hypothesis, for there may be other equally plausible hypotheses (see page 17). Support for any particular specific hypothesis can only be obtained by evaluating it against alternative hypotheses by analysis, modeling, and/or experimentation (see section 1.4).

(2) *Inter-population (trophic) mechanisms.* Delayed negative feedback can also be created by the interaction between populations of exploiter and exploited organisms (in this case A in figure 1.3 could be the number or density of exploiters and B the number or density of the exploited population). This usually means interactions between consumers ($A =$ abundance of herbivores or carnivores) and their resources ($B =$ abundance of plant or animal food). However, it is also possible for other exploitative interactions, such as mimicry, to create the necessary negative feedback architecture. Once again, plausible consumer–resource hypotheses can be erected if one can demonstrate a strong positive effect of the resource population on the consumer and a strong negative effect of the consumer on its resource (the reverse situation is unlikely to occur in this case). Before leaving this subject, it is necessary to point out that other, more complex, food web architectures can create the necessary conditions for endogenous population cycles. For example, if an interaction arrow was drawn from variable (species) C to A in figure 1.3, we would obtain a feedback loop involving three species; that is $A \to B \to C \to A$. Although third-order (or higher) feedback loops will have long time delays and, therefore, be capable of generating cyclic dynamics, they may not have the necessary negative sign. In order to generate plausible hypotheses for such structures, it is necessary to demonstrate strong interactions between all components of the loop, and that the overall sign of the loop (product of the interactions) is negative. It is important to realize that the strength of a feedback loop is the product of all the interactions in the loop, so that one weak interaction weakens the whole loop. This may be one reason that high-order dynamics are rarely detected in ecological systems.

1.4 How Can We Determine the Causes of Population Cycles?

This question may have been asked by more ecologists, more frequently, than any other single question. It has generated a multitude of papers and books, but no clear and unambiguous answer has emerged. This is partly due to philosophical and methodological issues concerning the analysis of circular causal systems like (1.2) (see also chapter 10).

Before beginning it is necessary to place the theoretical discussion (above) in a more realistic setting. In nature, populations are embedded within a web of interactions with elements of their physical and biotic environments.

Within this web may be many negative feedback loops capable of generating endogenous cycles; for example, interactions with food supplies, predators, parasitoids, pathogens, maternal effects, genetic effects, and so on. In addition, some environmental variables may themselves be cyclic (not due to interaction with the subject population), and may thereby drive exogenous cycles in the population of interest. The problem is how to untangle this ecological maze and recognize the real cause(s) of the observed dynamics. There are three basic approaches (as discussed below), all of which depend on the idea that only one, or at most a few, strong interactions (Paine 1980, 1992) or dominant feedback loops (Berryman 1993, 1999b) are responsible for generating the observed oscillations. The approaches discussed are designed to detect these strong interactions and dominant feedback loops in complex ecological systems.

1.4.1 Diagnosis

Diagnostic approaches employ statistical or other data analysis procedures (probes) to obtain information (clues) about the kind of oscillations being observed, and the mechanisms (sometimes called the deterministic skeleton) that generated them. Most of the statistical methods come from what is generally known as time series analysis.

Population data often consist of a time series—a series of observations on the abundance of an organism made at equal intervals (usually annually) over a fairly long period of time (e.g., figure 1.1). This type of data is commonly collected by agencies that consistently monitor resource or pest populations for management or conservation purposes. Time series analysis is designed to extract clues to the underlying generating structure and/or processes from such data.

1.4.1.1 Correlation within Time Series

Classical time series analysis (Box and Jenkins 1976) is based on the fact that, if a time series has a periodic component, there will be a strong positive correlation between observations (usually expressed as logarithms) separated by the length of the period (the lag), as shown in the autocorrelation function (ACF) (figure 1.4). Notice the large positive autocorrelation at lag 2 in the sycamore aphid series, indicating a regular oscillation with a period of 2 years, and at lag 9 in the budmoth series, indicating a 9-year cycle. The significance of the correlation can be roughly assessed by Bartlett's criterion, $2/\sqrt{n}$, where n is the length of the series. The ACF can also provide information on the general cause of the oscillation: If the cycle was generated by an exogenous factor, then the amplitude or height of the ACF should remain roughly constant as the lag gets larger, while if it decays with increasing lag, as it does in both the sycamore aphid and larch budmoth (figure 1.4), then the causal process is likely endogenous (Nisbet and Gurney 1982). Apart from

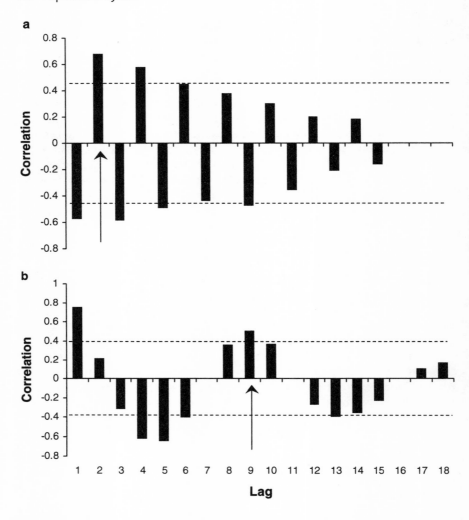

Figure 1.4 Autocorrelation functions for the sycamore aphid (a) and larch budmoth (b).

this, however, the ACF cannot tell us much about the feedback structure regulating population dynamics.

Endogenous population cycles are the result of second-order (or higher) feedback processes. The classical way to detect the order of a time series is to calculate the partial autocorrelation function (PACF) (Box and Jenkins 1976, Royama 1992). In biological systems, however, it makes more sense to calculate the correlation between lagged population density and the realized per-capita rate of change (Berryman 1999b, Berryman and Turchin 2001), to construct what is called a partial rate correlation function (PRCF) (figure 1.5). Because the presence of negative feedback always gives rise to a negative PRCF, only negative correlation coefficients need to be considered. Notice that sycamore aphid dynamics are dominated by first-order negative feedback (large negative correlation at lag 1), while larch budmoth dynamics are dominated by second-order feedback (large negative correlation at lag 2). Thus, PRCF informs us that larch budmoth population cycles could be caused by the mutual interaction between two dynamic variables. However, it cannot tell us much about the specific mechanisms, or even if these could involve intra- or interpopulation processes. To do this requires additional data.

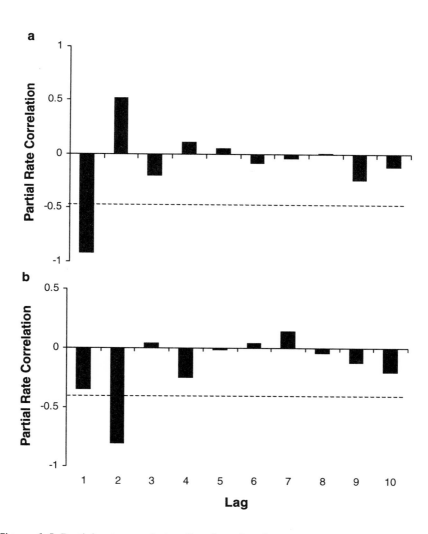

Figure 1.5 Partial rate correlation functions for the sycamore aphid (a) and larch budmoth (b).

1.4.1.2 Correlation between Time Series

Both exogenous and endogenous explanations for population cycles involve associations between two or more cyclic variables (see figures 1.2 and 1.3). For this reason we would expect to find a strong correlation between the causal variable and the response variable. This realization has led some investigators to embark on haphazard searches for correlation between arbitrarily chosen variables and the time series of interest (called "fishing trips" by some). The problem with this approach is that it often leads to spurious correlations and meaningless explanations (Yule 1926, Royama 1992). Hence, correlation analysis is not an end in itself, and should only be used to evaluate associations between variables that are known, or suspected, to affect each other. In the case of the larch budmoth, for example, larvae collected from samples were reared in the laboratory to determine the number containing insect parasitoids (figure 1.6a). Knowing that the interaction between predator and prey populations can create delayed negative feedback, the necessary condition for population cycles, it seems reasonable to examine the correlation between budmoth and parasitoid time series. At least finding a weak correlation may help reject a causal hypothesis (but see section 1.4 for an exception). The actual correlation between the budmoth and parasitoid time series is 0.938, which suggests that the two variables are closely associated. However, it tells us little about the causal process, as a strong correlation could be obtained from either an endogenous (the parasitoid–host interaction) or an exogenous (parasitoids tracking a budmoth cycle) structure.

Because changes in population numbers are manifest through the processes of individual reproduction and survival, it makes sense to examine the relationship between the individual rate of change of the population and the suspected causal variable(s), what I like to call the R-function (Berryman 1999b, 2001). In other words, we should examine the correlation between the realized per-capita rate of change of the larch budmoth and the density of parasitoids, or its natural logarithm (figure 1.7a). When we do this, however, we find that parasitism only explain about 21% of the variation in budmoth rates of change.

One problem with the above analysis is that it assumes a linear relationship between host survival and parasitoid density. However, there are good theoretical reasons to suspect that survival from parasitoid attack should be related to the percentage of parasitism, or the ratio of parasitoids to their prey (see section 1.4.2.1 for a discussion of ratio-dependent logistic models). In fact, the parasitoid/host ratio explains about 70% of the variation in budmoth rates of change (figure 1.7b).

Finding a strong relationship between a causal variable and the rate of change of the subject population still cannot tell us whether the dynamics are the result of an exogenous or an endogenous process. However, if the mechanism is endogenous, there should also be a strong correlation between the rate of change of the causal variable (in our case parasitoids) and the response variable (in our case the density of budmoth larvae or its logarithm).

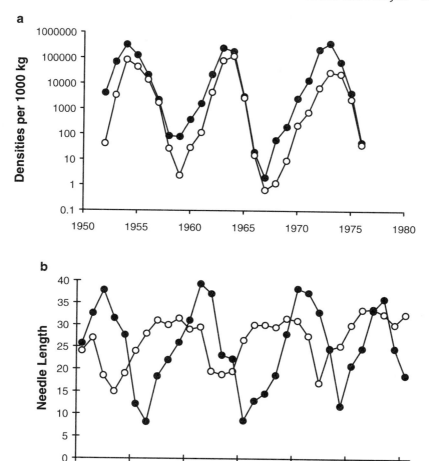

Figure 1.6 (a) Numbers of larch budmoth larvae (●) and those containing insect parasitoids (○) at Oberengadin and (b) numbers of budmoth larvae (●) at Sils compared with needle length (○) (data from W. Baltensweiler via P. Turchin).

In other words, we need to show that there is feedback between the two variables (e.g., as between A and B in figure 1.3). Again, the ratio-dependent relationship gives the best description of the interaction, explaining about 80% of the variability in the parasitoid rate of change (figure 1.8). Thus, correlation analysis of the respective R-functions suggests that there is strong feedback between budmoth and parasitoid populations, at least under the assumption of ratio-dependent interaction, and suggests that the observed cycles may be due to this second-order feedback structure. However, we still do not know if the feedback loop has the necessary negative sign.

16 Population Cycles

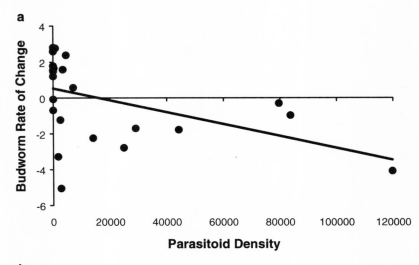

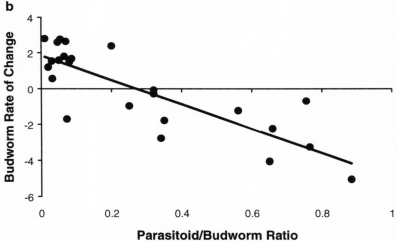

Figure 1.7 Relationships between the per-capita rate of change of the larch budmoth and the density of larvae containing insect parasitoids (a) and the parasitoid/host ratio (b).

The qualitative structure (or sign) of a feedback loop is obtained from the product of its individual interactions. In population systems, the qualitative effect of one population on another can be seen from the partial derivative of the R-function of the affected species evaluated with respect to the population affecting it. The feedback loop is negative if the product of all interactions (partial derivatives) is negative. In the case of the larch budmoth and parasitoid R-functions, it is fairly straightforward to show that this condition is met. This, together with the high coefficients of determination (70% and

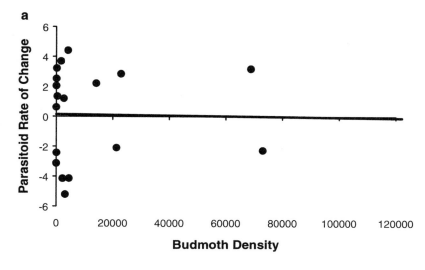

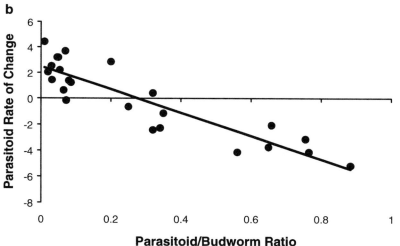

Figure 1.8 Relationships between the per-capita rate of change of parasitoids and the density of larch budmoth larvae (a) and the parasitoid/host ratio (b).

80%), suggests that the endogenous interaction between budmoths and insect parasitoids is a likely candidate for the delayed negative feedback process responsible for the observed population cycles. However, this should not dissuade us from examining other possibilities. For instance, one alternative hypothesis is that budmoth cycles are caused by interactions with its food supply (Fischlin and Baltensweiler 1979). According to this explanation, larch trees produce shorter, tougher, and more resinous needles in the years after heavy defoliation, and budmoth larvae feeding on these needles have lower survival and fertility rates (Benz 1974, chapter 7). Fortunately, we can examine the feedback between the budmoth population and its food supply because the length of larch needles was measured on one of the sampling plots although, alas, not the same one from which the parasitoid data came (figure 1.6b). In this case, however, the correlation between the two time series was extremely weak ($r = -.05$). In addition, needle length explained only about 24% of the variation in budmoth rates of change, while budmoth density explained 42% of the variability in needle length R-values, suggesting relatively weak feedback between the two variables. Although the derivatives of the R-functions satisfied the condition for negative feedback, the relatively weak interaction (in comparison with the stronger parasitoid–host interaction) suggests that it is not the main feedback loop involved in the cyclic oscillation. It could have a subsidiary, or modifying, effect on the budmoth cycle, however (see chapter 7).

It is interesting to note that, although we found signs of feedback between the budmoth population and larch needle lengths, the correlation between the two time series was negligible ($r = -.05$). This emphasizes the point that random searches for correlation between time series are unlikely to uncover the real causes of population cycles, and that they may, in fact, completely mislead the analysis.

Of course, it would be ideal if all the factors suspected of affecting the dynamics of a given population were measured, for then one could examine the relative impact of each factor singly and in combination (multivariate analysis). Unfortunately, this is rarely done (however, see chapter 2). Plans to monitor biological populations in the future should seriously consider the obvious benefits of collecting multiple, coincident, time series data (co-variates).

1.4.2 Models

Models can take many forms, ranging from extremely complicated assemblages of biological and ecological facts to simple theoretical models that concentrate on the basic ecological principles and/or processes. The former approach was popular in the middle of the last century, but has generally fallen out of favor because the models were too complicated and opaque (i.e., the models were almost as complicated and difficult to understand as the real system). The tendency nowadays is to use a more theoretical approach because the models so produced are more readily understood and interpreted.

1.4.2.1 Time Series Models

Time series models are usually developed as forecasting tools. However, they are also useful for evaluating the dynamic behavior of modeled populations to see how well they simulate the natural system. Comparing the behavior of different models may also help us to choose between competing models and their associated hypotheses.

Classical time series models (Box and Jenkins 1976) have the general form

$$\ln N_t = a + b \ln N_{t-1} + c \ln N_{t-2} + \cdots + k \ln N_{t-d}, \tag{1.3}$$

where N_t is the value of the measured variable (say the density of the budmoth population) at time t, d is the maximum delay (or order) of the system, and $a, b, c, \ldots, k$ are parameters estimated by linear regression. However, as discussed earlier, it makes more sense to model the per-capita rate of change (the R-function) of biological populations (Royama 1992, Berryman 1999b); that is,

$$R_t = a + b \ln N_{t-1} + c \ln N_{t-2} + \cdots + k \ln N_{t-d}, \tag{1.4}$$

where $R_t = \ln N_t - \ln N_{t-1}$ is the realized per-capita rate of change over a unit period of time. On theoretical grounds, some authors (including this one) prefer not to use logarithms on the right-hand side of the equation (e.g., Berryman 1999b, Berryman and Turchin 2001), in which case equation (1.4) becomes

$$R_t = a + bN_{t-1} + cN_{t-2} + \cdots + kN_{t-d} \tag{1.5}$$

The model can be made more flexible by adding exponents to the variables, as in the "response surface methodology" employed by Turchin and Taylor (1992).

Remembering that time delays can be created by reciprocal interactions with other elements in the environment, a corresponding model for multiple time series data is

$$R_t^N = a^N + b^N N_{t-1} + c^N P_{t-1} + \cdots + d^N F_{t-1}, \tag{1.6}$$

where R^N is the realized per-capita rate of change of one population (say budmoths), and $P, \ldots, F$ represent the values of other variables (say parasitoids and foliage quality). (Notice that this equation does not contain any time delays greater than one.) To employ this model we also need time series data for the other variables, so we can write equations for their R-functions. For example, the budmoth and parasitoid data can be modeled by a system of two equations

$$\begin{aligned} R_t^N &= a^N + b^N N_{t-1} + c^N P_{t-1}, \\ R_t^P &= a^P + b^P P_{t-1} + c^P N_{t-1}. \end{aligned} \tag{1.7}$$

Of course, one could use logarithms and/or place exponents on the variables on the right-hand side of this equation if so desired.

Models in which the per-capita rate of change is linearly related to the variables [as in the system (1.7)] belong to the well-known Lotka–Volterra (LV) family. If logarithms are used on the right-hand side, then they become Gompertz modifications of the LV system (GLV). However, there may be good theoretical reasons to expect the rate of change to depend on the consumer/resource ratio, in which case logistic models should describe the data better (Berryman 1999a,b, Berryman and Gutierrez 1999). A ratio-dependent logistic model for the budworm–parasitoid system would have the form

$$R_t^N = a^N + b^N N_{t-1} + c^N P_{t-1}/(w + N_{t-1}),$$
$$R_t^P = a^P + b^P P_{t-1} + c^P P_{t-1}/(w + N_{t-1}).$$
(1.8)

where w represents the relative availability of alternative prey for the parasitoids (if alternative prey species are measured, w can be replaced by their weighted sum, otherwise w is normally assumed to be zero or estimated by convergence methods after linear regression).

Parameters of all the above models can be estimated by linear regression of the realized per-capita rate of change of each species on the variables on the right-hand side of the equations. The choice of a particular model is usually based on some "goodness of fit" criterion (e.g., coefficients of determination), unless there is some rational reason to choose one model over another. In the case of the budmoth and parasitoid time series, the logistic model fits better than the others (parameters are given in figure 1.9a). In this figure, the larch budmoth dynamics predicted by the fitted logistic model are compared with the actual time series data. Notice that the model simulates the dynamics fairly well for the first two cycles, but then begins to depart from the real series. In fact, the amplitude of the modeled trajectory slowly declines with time to a stable cycle of period 8 years (compared with the real 9-year cycle) and a 150-fold amplitude (compared with the real 10,000-fold amplitude). If a small amount of random variability is added to the computed values of R, however, the amplitude and period of the cycles can be sustained indefinitely. Despite these encouraging signs, there are some problems in the details of the simulated dynamics. For example, the model predicts an increase phase of 5–6 years and a decrease phase of 2–3 years, while examination of a number of real data sets (from Baltensweiler 1989 and elsewhere) indicates an actual increase phase of 5.44 ± 0.73 years and a decrease phase of 3.69 ± 0.75 years.

Figure 1.9b shows the dynamics of the larch budmoth population predicted by a LV time series model of the interaction between larch budmoth and needle lengths (food quality). In this case the model damps quickly to a stable equilibrium and random variability fails to recreate cycles of the correct period and amplitude. It seems unlikely, therefore, that this interaction alone could be responsible for the observed population cycles.

What do these results tell us about the causes of population cycles in the larch budmoth? Diagnostic analysis indicates that the interaction with insect parasitoids explains much more of the variation in larch budmoth dynamics than needle length, and the parasitoid model simulates the dynamics more closely than a needle-length model. This may lead us to conclude that endogenous feedback between budworm and parasitoid populations is more likely to be the cause of the observed population cycle. However, there are enough inconsistencies between simulated and observed dynamics to suggest that other factors, like food quality, could play a subsidiary role. It is indeed unfortunate that parasitoids and food quality were not measured on the

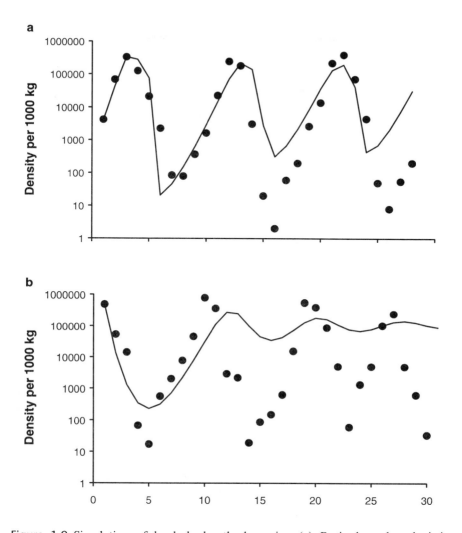

Figure 1.9 Simulation of larch budmoth dynamics. (a) Ratio-dependent logistic model of the host–parasitoid interaction [equation (1.8)] with parameters estimated from the budmoth and parasitoid time series (figure 1.6a); parameter values for the budmoth (N) R-function are $a^N = 2.455$, $b^N = 0.0000074$, $c^N = 7086$ ($r = .91$) and for the parasitoids (P) R-function $a^P = 2.575$, $b^P = 0$, $c^P = 8.83$ ($r = .9$) with alternative prey $w = 0$. (b) Lotka–Volterra model of the food–budworm interaction [equation (1.7)] with parameters estimated from the budmoth and needle-length time series (figure 1.6b); parameter values for the needle-length (N) R-function are $a^N = 0.515$, $b^N = -0.162$, $c^N = -0.00000063$ ($r = .7$) and for the budmoth (P) R-function $a^P = -9.048$, $b^P = -0.000003$, $c^P = 0.361$ ($r = .55$). Simulations begin with the first two data points (N_1, P_1). The values of R for each population are calculated from the relevant R-function [equation (1.7) or (1.8)], and then the new variable (N_2, P_2) is calculated from the exponential growth equation $X_t = X_{t-1} \exp(R^x)$, $X = N; P$. This procedure is repeated for as many time steps as are needed to simulate the dynamic trajectories.

same plots so that their combined effect on budworm dynamics could be assessed. In the absence of such data, the only solution is to build a hypothetical model of the total interactive system to see if it can improve the predictions, or perform definitive experiments (see section 1.4.3).

1.4.2.2 Hypothetical Models

Hypothetical models attempt to capture, in a fairly realistic manner, the specific processes that are thought (hypothesized) to give rise to a particular dynamic pattern (some call these "mechanistic" models). In other words, a hypothesis is formulated about the causal structure (often based on diagnostic analysis), a model is built to describe this hypothesized structure, parameters of the model are estimated from available data, and the output of the model is compared with the real data. Most hypothetical consumer–resource models are built around the functional response of the consumer, from which comes the death rate of the prey and the birth rate of the predator. For example, Hanski and Henttonen (chapter 3) show how a hypothetical model for the interactions between vole populations and their specialist and generalist predators can describe the cyclic populations in northern Fennoscandia and noncyclic populations in the south.

When time series are available for both consumer and resource populations, parameters can be estimated directly from the data. In chapter 7, Turchin and his colleagues develop a hypothetical model of the larch budmoth population system, with some parameters estimated from the time series, to demonstrate how the combined interaction with food and parasitoids can describe the details of the budmoth cycle.

Hypothetical models are particularly useful for testing ideas about the causes of population cycles to see if they can, at least in theory, describe the dynamics of the real system and/or experimental results. They are a useful prelude to field experiments, as they provide plausible hypotheses and testable predictions (see, for example, chapters 5, 6, and 7).

1.4.2.3 Systems Models

Systems models attempt to describe all the factors and interactions affecting the dynamics of a given population. Once this has been done, the variable(s) and feedback loop(s) involved in generating the observed dynamics can often be determined by a process of elimination (see chapter 2, for example).

There are many ways of building systems models, but probably the most popular evolves from the life-table approach (Varley et al. 1973). Here, survival from one age class or stage of development to the next is explicitly modeled as a function of predators, parasitoids, pathogens, and other mortality agents. Submodels are then built that describe the reproduction and survival of the various agents of mortality, and these are joined together to form a complete model of the interactive system. Provided the model behaves in a similar manner to the real system, it then becomes possible to perform experiments on the surrogate (modeled) system. For example, we could set a particular variable constant to see what effect it has on the dynamics of the model. We could also set constant all other variables, except the one being evaluated. By this process it may be possible to home in on the variable(s) and feedback loop(s) responsible for generating cycles of the correct amplitude and period (see chapter 2).

1.4.3 Experiments

Diagnosis and modeling provide useful information, or inferences, about the probable causes of observed population oscillations, and lead to the development of plausible strong hypotheses. The obvious next step is to test these hypotheses experimentally. On occasion, the analytical inferences may be so strong that it is difficult to convince people to perform experiments. An example is the spruce needleminer (chapter 2), where the analytical case for parasitoid–host feedback is so compelling that nobody seems willing to perform the definitive experiment, even though it would be fairly easy and inexpensive to do. In other cases experiments may be impossible for practical or financial reasons. For example, population studies are often sponsored by resource or pest management agencies, which rarely have the patience or finances to carry out long-term, large-scale, expensive, field experiments. A possible solution to this problem is to employ an "adaptive" management strategy, in which normal management operations, such as different harvesting intensities, can double as experiments (Holling 1978).

Even when they are practically and financially feasible, the design of experiments capable of distinguishing between competing hypotheses under complex field conditions is no simple task. Most experimental designs involve the removal or reduction of the hypothesized causal variable(s), say by excluding predators with cages (chapters 4 and 5) or removing parasitic infections with antibiotics (chapter 6). However, even if they are successful (say by stopping the cycle) they may not solve the problem completely. For instance, suppose we want to explain the causes of cycles in population C (figure 1.3) by conducting an experiment that keeps the causal variable B constant. If this treatment suppresses cycles of C it would tend to eliminate an intrapopulation explanation (i.e., H5 or H6), but it could not distinguish between an exogenous cycle (C following B) and an endogenous cycle (mutual interaction between B and C). To prove the point requires the reverse experiment, keeping C constant. If B still cycles, then the cause of the cycle in C must be exogenous, while if B ceases to cycle, then the cause must be endogenous. Another problem involves the notion of feedback hierarchies (Berryman et al. 1987), which suggests that, even if an experiment removes the causal feedback loop, another could immediately replace it, causing the population to continue cycling and the experimenter to reject the hypothesis when it was, in fact, true. Even when an experiment is successful, or partly successful, it is rarely the end of the story, as illustrated by the continuing debate over the causes of the grouse cycle in Britain (see chapters 6 and 9). There may be many different ways to interpret the results of any given experiment.

Perhaps the best way to design critical experiments is to use models to formulate testable hypotheses. For example, Turchin et al. (1991) employed hypothetical models to develop a series of predictions about what would happen if sections of a tree were caged to exclude predators during a southern pine beetle outbreak. They then went on to test these predictions and came to the conclusion that the results were consistent with the theory that predators had a delayed negative feedback effect on southern pine beetle dynamics (chapter 5). Of course, demonstrating a delayed negative feedback effect of predators is not sufficient to prove the point, for there may be other stronger exogenous or endogenous processes at work.

1.5 Summary

In this chapter, I have attempted to provide a conceptual and analytical framework for the study of cyclic animal populations. I first defined a cyclic population as one that exhibits "an oscillation in population numbers or density that has an obviously regular period of three or more years." In general terms, an oscillation of this kind can be caused by an exogenous variable that oscillates with the same frequency, or by an endogenous second-order (or higher) negative feedback loop. In the latter case, the delayed negative feedback can be caused by intrapopulation processes, such as maternal or genetic effects, or by interpopulation processes, like predator–prey or plant–herbivore (trophic) interactions.

Analytical methods are intended to discover the endogenous or exogenous processes that are responsible for destabilizing the system and generating the observed cyclical dynamics. When time series data are available for only one species, autocorrelation analysis can provide clues to the period and order of the dynamics, and whether exogenous or endogenous forces are responsible. When time series are also available for other populations (e.g., food, predators, pathogens, parasitoids), multiple regression analysis of the per-capita rate of change against the densities (or ratios) of the other populations can be used to evaluate their contribution to the dynamics of the subject species. The presence of a feedback loop can then be evaluated by reverse regression. Strong negative feedback, the necessary condition for endogenous cycles, is suggested by high coefficients of determination both ways, and a negative sign for the feedback loop. Models fitted to time series data or built around hypothesized relationships can then be analyzed to see which best reconstructs the observed oscillations. In some cases it may be possible to build complete systems models, which allow one to evaluate the contribution of various factors by the process of elimination. The analysis of time series data and models leads to strong plausible hypotheses about the causes of population cycles, and points the way to definitive experiments capable of distinguishing between competing hypotheses.

ACKNOWLEDGMENTS AND COMMENTS

I am extremely grateful to Tony Dixon, Werner Baltensweiler, and Peter Turchin for supplying me with time series data, and to Xavier Lambin, Matt Ayres, Mauricio Lima, Fred Wagner, Ilkka Hanski, and Charley Krebs for providing thoughtful suggestions to improve this chapter. In my subsequent efforts, I have tried to incorporate most of their suggestions. What is left are our disagreements, the most serious of which seems to be my apparent overemphasis of time series analysis at the expense of the experimental approach (see also chapter 9). Perhaps I need to explain this. The experimental approach has dominated biology for more than a century but has, strangely, not solved the problem of population cycles. Why is this? The first reason may be one of scale, as experiments aimed at answering population-level questions need to be performed at the appropriate spatial scale. But what is the correct spatial identity of "a population." Nobody really seems to know (although some of us are grappling with this problem; e.g., Berryman 2002, Camus and Lima 2002), but most seem to agree that whatever it is, it is large (on the order of hectares) rather than small (on the order of meters). This leads to problems of design and logistics: How big an area do we need to treat and for how long do we need to treat it (e.g., by predator exclusion) to test population-level hypotheses. This is further exacerbated by the fact that many of the predators involved move freely over vast areas. A second reason may be philosophical. The experimental approach evolved during the time when unidirectional causal thinking [e.g., process (1.1)] was the scientific modus operandus. What this means is that a simple experiment could usually be designed to expose the unidirectional causal relationship between two variables. However, the identification of feedback processes [e.g., process (1.2)] may require two experiments. In addition, the fact that a variable involved in a feedback loop affects *itself* after a certain (often unknown) period of time makes timing critical and interpretation difficult. In contrast, ecological time series analysis is a recent development that is not widely used or appreciated in ecological circles, where the experimental approach holds sway. Yet it offers a powerful way to look at both unidirectional and mutually causal systems. This is why I tend to emphasize and encourage the development and use of time series analysis. However, I still believe, like most ecologists, that experiments will have the final say on the problem of population cycles, although it may well be diagnosis and models that identify the definitive experiment(s).

REFERENCES

Anderson, R. M. and R. M. May. 1980. Infectious diseases and population cycles of forest insects. *Science* 210: 658–661.

Baltensweiler, W. 1978. Ursache oder Wirkung? Huhn oder Ei? *Bull. Soc. Entomol. Suisse* 51: 261–267.

Baltensweiler, W. 1989. The folivore guild on larch (*Larix decidua*) in the Alps. In Y. N. Baranchikov, W. J. Mattson, F. P. Hain, and T. L. Payne (Eds.) *Forest insect guilds: patterns of interactions with host trees*. USDA Forest Service General Technical Report NE-153, pp. 145–164.

Benz, G. 1974. Negative Ruckkoppelung durch Raum- und Nahrungskonkurrencz sowie sykliche Veranderung der Narungsgrundlage als Regelprinzip in der Populationsdynamik des Grauen Larchensicklers, *Zeiraphera diniana* (Guenee) (Lep. Tortricidae). *Z. Angew. Entomol.* 76: 31–49.

Berryman, A. A. 1973. Population dynamics of the fir engraver *Scolytus ventralis* (Coleoptera: Scolytidae). I. Analysis of population behavior and survival from 1964 to 1971. *Can. Entomol.* 105: 1465–1488.

Berryman, A. A. 1981. *Population systems: a general introduction.* Plenum Press, New York.

Berryman, A. A. 1986. On the dynamics of blackheaded budworm populations. *Can. Entomol.* 118: 775–779.

Berryman, A. A. 1989. The conceptual foundations of ecological dynamics. *Bull. Ecol. Soc. Am.* 70: 234–240.

Berryman, A. A. 1993. Food web connectance and feedback dominance, or does everything really depend on everything else? *Oikos* 68: 183–185.

Berryman, A. A. 1999a. Alternative perspectives on consumer–resource dynamics: a reply to Ginzburg. *J. Anim. Ecol.* 68: 1263–1266.

Berryman, A. A. 1999b. *Principles of population dynamics and their application.* Stanley Thornes, Cheltenham, UK.

Berryman, A. A. 2001. Functional web analysis: detecting the structure of population dynamics from multi-species time series. *Basic Appl. Ecol.* 2: 311–321.

Berryman, A. A. 2002. Population: a central concept for ecology? *Oikos* 97: 439–442.

Berryman, A. A. and A. P. Gutierrez. 1999. Dynamics of insect predator–prey interactions. In C. B. Huffaker and A. P. Gutierrez (Eds.) *Ecological entomology.* John Wiley, New York, pp. 389–423.

Berryman, A. A. and P. Turchin. 2001. Identifying the density-dependent structure underlying ecological time series. *Oikos* 92: 265–270.

Berryman, A. A., N. C. Stenseth, and A. S. Isaev. 1987. Natural regulation of herbivorous forest insect populations. *Oecologia* 71: 174–184.

Bjørnstad, O. N., W. Falck, and N. C. Stenseth. 1995. A geographic gradient in small rodent density fluctuations: a statistical modelling approach. *Proc. Roy. Soc. Lond., Ser. B* 262: 127–133.

Bodenheimer, F. 1938. *Problems of animal ecology.* Oxford University Press, London.

Box, G. E. P. and G. M. Jenkins. 1976. *Time series analysis: forecasting and control.* Holden Day, Oakland, Calif.

Camus, P. A. and M. Lima. 2002. Populations, metapopulations, and the open–closed dilemma: the conflict between operational and natural population concepts. *Oikos* 97: 433–438.

Chitty, D. 1957. Self-regulation of numbers through changes in viability. *Cold Spring Harbor Symp. Quant. Biol.* 22: 277–280.

Chitty, C. 1967. The natural selection of self-regulatory behavior in animal populations. *Proc. Ecol. Soc. Austral.* 2: 51–78.

Chitty, D. 1996. *Do lemmings commit suicide? Beautiful hypotheses and ugly facts.* Oxford University Press, New York.

Christian, J. J. 1950. The adreno-pituitary system and population cycles in mammals. *J. Mammal.* 31: 247–259.

Dixon, A. F. G. 1990. Population dynamics and abundance of deciduous tree-dwelling aphids. In A. D. Watt, S. R. Leather, M. D. Hunter, and N. A. C. Kidd (Eds.) *Population dynamics of forest insects.* Intercept, Andover, UK, pp. 11–23.

Elton, C. S. 1924. Periodic fluctuations in the numbers of animals: their causes and effects. *Br. J. Exp. Biol.* 2: 119–163.

Elton, C. S. 1942. *Voles, mice and lemmings.* Clarendon Press, Oxford.

Finerty, J. P. 1980. *The population ecology of cycles in small mammals.* Yale University Press, New Haven, Conn.

Fischlin, A. and W. Baltensweiler. 1979. Systems analysis of the larch budmoth system. Part 1. The larch–larch budmoth relationship. *Bull. Soc. Entomol. Suisse* 52: 273–289.

Ginzburg, L. R. and D. E. Taneyhill. 1994. Population cycles of forest Lepidoptera: a maternal effect hypothesis. *J. Anim. Ecol.* 63: 79–92.

Haukioja, E. and T. Hakala. 1975. Herbivore cycles and periodic outbreaks: formulation of a general hypothesis. *Rep. Kevo Subarct. Res. Stat.* 12: 1–3.

Holling, C. S. (Ed.) 1978. *Adaptive environmental assessment and management.* John Wiley, New York.

Hunter, M. D. and P. W. Price. 1998. Cycles in insect populations: delayed density dependence or exogenous driving variables? *Ecol. Entomol.* 23: 216–222.

Hutchinson, G. E. 1948. Circular causal systems in ecology. *Proc. New York Acad. Sci.* 50: 221–246.

Keith, L. B. 1963. *Wildlife's ten-year cycle.* University of Wisconsin Press, Madison, Wisc.

Krebs, C. J. and J. H. Myers. 1974. Population cycles in small mammals. *Adv. Ecol. Res.* 8: 267–399.

Lotka, A. J. 1924. *Elements of mathematical biology* (republication of *Elements of physical biology*, first published in 1924). Dover, New York.

Milsum, J. H. 1966. *Biological control systems analysis.* McGraw-Hill, New York.

Morris, R. F. 1959. Single-factor analysis in population dynamics. *Ecology* 40: 580–588.

Moss, R. and A. Watson, 1991. Population cycles and kin selection in red grouse *Lagopus lagopus scoticus.* Ibis Suppl. 1: 113–120.

Myers, J. H. 1988. Can a general hypothesis explain population cycles of forest lepidoptera? *Adv. Ecol. Res.* 18: 179–242.

Nisbet, R. M. and W. S. C. Gurney. 1982. *Modelling fluctuating populations.* John Wiley, Chichester, UK.

Paine, R. T. 1980. Food-web: linkage, interaction strength and community infrastructure. *J. Anim. Ecol.* 49: 667–685.

Paine, R. T. 1992. Food-web analysis through field measurement of per-capita interaction strength. *Nature* 355: 73–75.

Pitelka, F. A. 1957. Some aspects of population structure in the short-term cycle of the brown lemming in northern Alaska. *Cold Spring Harbor Symp. Quant. Biol.* 22: 237–251.

Rossiter, M. C. 1991. Environmentally-based maternal effects: a hidden force in population dynamics? *Oecologia* 87: 288–294.

Royama, T. 1992. *Analytical population dynamics.* Chapman and Hall, London.

Stenseth, N. C. 1999. Population cycles in voles and lemmings: density dependence and phase dependence in a stochastic world. *Oikos* 87: 427–461.

Turchin, P. and A. A. Berryman. 2000. Detecting cycles and delayed density dependence: a comment on Hunter and Price (1998). *Ecol. Entomol.* 25: 119–121.

Turchin, P. and A. D. Taylor. 1992. Complex dynamics in ecological time series. *Ecology* 73: 289–305.

Turchin, P., P. L. Lorio, A. D. Taylor, and R. F. Billings. 1991. Why do populations of southern pine beetle (Coleoptera: Scolytidae) fluctuate. *Environ. Entomol.* 20: 401–409.

Varley, G. C., G. R. Gradwell, and M. P. Hassell. 1973. *Insect population ecology: an analytical approach.* Blackwell Scientific, Oxford.

Volterra, V. 1926. Fluctuations in abundance of a species considered mathematically. *Nature* 118: 558–560.
Wellington, W. G. 1960. Qualitative changes in natural populations during changes in abundance. *Can. J. Zool.* 38: 290–314.
White, T. C. R. 1974. A hypothesis to explain outbreaks of looper caterpillars, with special reference to populations of *Selidosema suavis* in a plantation of *Pinus radiata* in New Zealand. *Oecologia* 16: 279–301.
Witting, L. 1997. *A general theory of evolution by means of selection by density dependent competitive interactions*. Peregrine, Århus.
Wright, L. E., A. A. Berryman, and B. E. Wickman. 1984. Abundance of the fir engraver, *Scolytus ventralis*, and the Douglas-fir beetle, *Dendroctonus pseudotsugae*, following tree defoliation by the Douglas-fir tussock moth, *Orgyia pseudotsugata*. *Can. Entomol.* 116: 293–305.
Yule, G. U. 1926. Why do we sometimes get nonsense-correlation between time-series? A study in sampling and the nature of time-series. *J. Roy. Stat. Soc.* 89: 1–64.

2

The Role of Insect Parasitoids in Population Cycles of the Spruce Needleminer in Denmark

Mikael Münster-Swendsen

2.1 Introduction

The spruce needleminer, *Epinotia tedella* (Cl.) (Lepidoptera: Tortricidae), is a small and abundant moth associated with Norway spruce (*Picea abies* Karst.). Larvae mine spruce needles, usually those more than 1 year old, and each requires about 35 needles to meet its food demands. In central Europe, the spruce needleminer is regarded as a temporary, serious pest when densities reach several thousand per square meter. However, it seldom causes significant damage in Scandinavian countries. An exception was the heavy infestation in southern Denmark in 1960–61.

The spruce needleminer has one generation per year. Adults emerge in June and deposit eggs singly on spruce needles. Larvae mine the needles from July through October and then descend on silken threads in November to hibernate in the forest litter as prepupal larvae in cocoons. Pupation occurs in early May and lasts 3–4 weeks.

Like many other forest defoliators, spruce needleminers are associated with a diverse fauna of parasitic Hymenoptera (parasitoids) (Münster-Swendsen 1979). Eggs are attacked by a minute wasp (*Trichogramma* sp.) that kills the embryo and emerges as an adult a few weeks later. Because spruce needleminer eggs have all hatched by this time, the parasitoids must oviposit in the eggs of other insect species. In other words, this parasitoid is not host-specific and therefore not expected to show a numerical response to spruce needleminer population changes.

Newly hatched moth larvae immediately bore into needles and, because of this, are fairly well protected against weather and predators. However, specialized parasitic wasps (parasitoids) are able to deposit their eggs inside a larva by penetrating the needle with their ovipositor. Two species, *Apanteles tedellae* (Nix.) and *Pimplopterus dubius* (Hgn.), dominate the parasitoid guild and sometimes attack a large percentage of the larvae (Münster-Swendsen 1985). Parasitized larvae continue to feed and, in November, descend to the forest floor to overwinter with unparasitized individuals. In late April, however, the parasitoids take over and kill their hosts.

Besides mortality from endoparasitoids, up to 2% of the larvae die within the mine due to an ectoparasitoid and a predatory cecidomyid larva. Because they descend to the forest floor so late in the year, when most other insect activity has ceased, larvae suffer little mortality from November to March. In spring, some larvae appear to be infected with a pathogenic fungus [*Paecilomyces farinosus* (Holm.)], and these die and become "moulded mummies" in April. The remains of these "mummies" (sclerotia) may then infest the descending larvae of the following generation. In May, numerous larvae of a soil-dwelling click-beetle [*Athous subfuscus* (Müll.)] move to the upper soil layers where they attack and eat cocoons and pupae of all kinds of insects, including needleminer larvae and pupae.

Only needleminers that escape parasitism, fungus infection, and predation emerge as adults in June, but their condition may vary considerably. Some (up to 50%) are infected by a protozoan parasite (*Mattesia* sp.) that reduces their life-span and potential fecundity (Münster-Swendsen 1991). Others suffer from dramatically reduced gonad function without an apparent cause (but see section 2.4.1). Because of these factors, female needleminers rarely achieve their maximum fecundity of around 80 eggs each. In my study, the estimated average fecundity varied between 0.69 and 46.75 eggs/female over 13 years.

Field studies were initiated in Danish spruce stands in 1970. My main objectives were to develop life tables for the spruce needleminer, determine the key factors affecting their population dynamics, and build a detailed simulation model describing their interaction with the major environmental factors. More recently, analysis of the time series collected over 20 years has improved our understanding of the causes of the observed cyclic dynamics, as demonstrated in this chapter.

2.2 Census Methods and Data

Data were collected over the period 1970–89 by intercepting larvae with funnel-traps as they descended from the canopy during November. Captured individuals were taken to the laboratory for dissection and identification of parasitoids. In this way a 19-year record (time series) of needleminer density per unit area of forest floor, and the frequency of parasitism by each parasitoid species (three primary, two clepto-, and one hyperparasitoid species), was obtained. Sampling was carried out in 11 spruce stands but over differing periods of time. Altogether, more than 30,000 larvae were counted and dissected.

In order to establish detailed life tables, more intensive sampling was carried out over a 10-year period in one of the spruce stands. In this case all developmental stages were sampled throughout the year and the levels and causes of the mortalities recorded. As with many other univoltine insects, a fairly discrete shift occurred between various mortality factors during the year, and this made separation and quantification of each mortality factor more feasible. Samplings of branches in August and October revealed the percentages killed by an egg parasitoid (eggshells remain firmly attached to needles), a gall-midge larva, and an ectoparasitoid attacking the needleminer in its mine, while samples of the upper soil layer in April and May provided mortality rates due to the fungus that kills hibernating larvae, and soil predators like the abundant click-beetle larvae.

Absolute estimates of adult insects emerging per unit area of forest floor were obtained from 200 emergence traps placed on the ground from early June until mid-July. Emerging moths and parasitoids were identified and counted. These samples permitted estimation of total winter mortality and of realized fecundity, that is, density of eggs divided by density of adults, with egg density estimated by adding the density of descending larvae, egg mortality, and larval mortality.

In all, a total of 92 stand-years of data were accumulated, including complete life tables for about 10 stand-years. The sequence of mortality causes and their percentage ranges are shown in table 2.1 (for further details on sampling methods and data, see Münster-Swendsen 1985).

A synoptic overview of needleminer population fluctuations in Denmark was obtained by combining data from my studies in North Zealand (1970–89), a countrywide survey (1989–93), and reports from foresters (1960–69) (figure 2.1). Since fluctuations are synchronous across stands within a larger region (figure 2.2), the 1970–93 data show averages across all sampled stands. The most conspicuous feature of this time series is the more-or-less regular 6–7-year cycle in needleminer abundance. Density changes of needleminer and parasitoid larvae over 19 years in one spruce stand are shown in figure 2.3, clearly illustrating their coherent cycles and suggesting a significant role for parasitoids in the population dynamics of the needleminer.

Table 2.1 Life-table sequence of mortalities and fecundity in *Epinotia tedella*

Stage	Cause	Range (%)
Egg	*Trichogramma*	7–26
Larva (young)	Endoparasitoids	6–86
Larva (medium)	Ectoparasitoid	0–2
Larva (medium)	Predator (Cecidomyid)	0–2
Larva (hibernating)	Fungus disease	15–44
Pupa	Predator	20–42
Adult (fec. red.)	Pseudoparasitism	10–95
Adult (fec. red.)	Protozoan infection	< 50

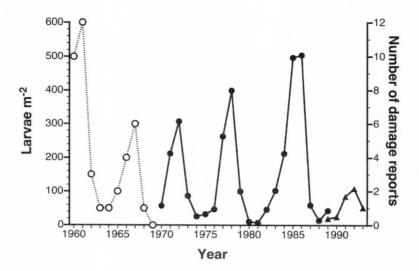

Figure 2.1 Densities of *E. tedella* larvae from 1970–89 (average of 1–7 spruce stands of 2–22 ha each in North Zealand) and from 1989–93 (average of 12 stands of 1/4 ha each spread all over Denmark). The period 1960–69 is represented by the number of damage reports submitted by Danish foresters.

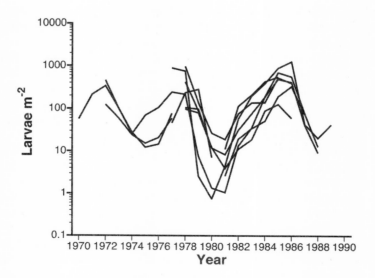

Figure 2.2 Synchronous fluctuations of needleminer populations in 11 isolated spruce stands within a region of 70 km^2 in North Zealand.

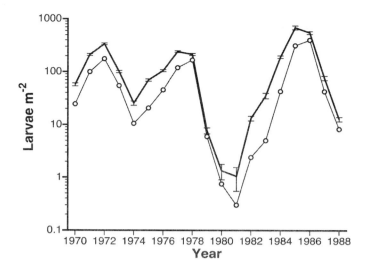

Figure 2.3 Densities (±SE) of *E. tedella* larvae descending from their host trees in November within one stand (thick line) and of parasitized larvae (thin line).

2.3 Life Table Analysis

Life table and key factor analyses are traditionally used to identify the important factors acting on population dynamics (Varley et al. 1973). Using this approach, I found that mortalities due to egg parasitoids and soil predators were independent of population density, and that larval mortality within the mine was negligible (Münster-Swendsen 1989). Mortality due to mycosis (fungus disease) showed a low, and hardly detectable, relationship to larval density in the previous year, whereas there was a significant relationship to larval parasitism, suggesting the possibility of delayed negative feedback (figure 2.4). The primary key factor was found to be reduction in fecundity (maximum fecundity minus observed fecundity), which also showed evidence of delayed negative feedback on population density (figure 2.5).

The importance of fecundity reduction stimulated an intense search for the causes of reproductive malfunction. Several thousand adults were dissected and their organs and hemocoel examined. Aside from variation in condition of the ovaries, with no apparent cause, spores of a protozoan (*Mattesia* sp., Neogregarinidae) were found in some adults. When spores were numerous, female ovaries tended to be shrunken and adults had shorter life-spans (Münster-Swendsen 1991). However, during declines of needleminer populations, fewer than 60% of the adults were infected and the majority of these contained relatively few spores. As a result, protozoan infections could not explain the 90% reduction in fecundity often observed during population declines.

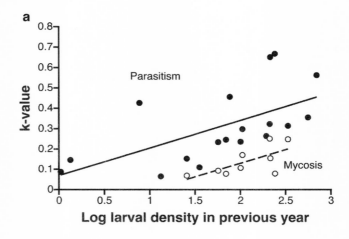

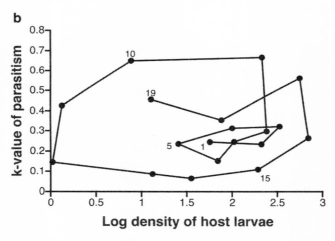

Figure 2.4 (a) The effect of population density in the previous year on parasitism (●) and mycosis (○); parasitism $n = 18$, $r^2 = .37$, $P < .0005$; mycosis $n = 9$, $r^2 = .45$, $P < .025$. (b) Mortalities (k-values) due to parasitism plotted against host densities and connected in temporal sequence. The anticlockwise, eliptical pattern indicates delayed negative feedback.

2.3.1 Influence of Weather

Analysis of weather patterns over the period 1970–89 failed to detect any weather factors (temperatures and precipitation) that could explain the cyclic fluctuations in needleminer populations, nor the observed variation in fecundity. However, the physiological condition of spruce trees in Denmark (expressed by radial growth and a vigor index) was found to be strongly influenced by May–August precipitation, and the average dry weight of descending larvae was negatively correlated with summer precipitation

The Role of Insect Parasitoids 35

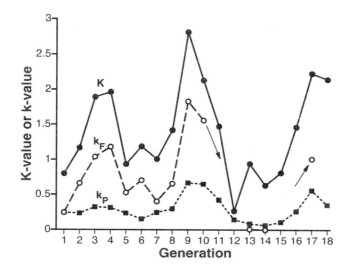

Figure 2.5 Graphical key factor demonstration. Generation mortality, K (●), parasitism, k_P (■), and fecundity reduction, k_F (○), are plotted in temporal sequences. The factor that mimics the pattern of fluctuations and amplitudes of the generation mortality is identified as the key factor; that is, the factor that contributes most significantly to the population fluctuations.

(Münster-Swendsen 1984). This suggested that there may be a negative correlation between adult fecundity and precipitation in May–August of the preceding year, but neither this nor a delayed effect of precipitation on population change could be statistically demonstrated (Münster-Swendsen in preparation). The influence of weather on the dynamics of the spruce needleminer may be of a more complex nature and is apparently overwhelmed by biotic factors.

2.3.2 Simulation Model and Predictions

A complex simulation model was constructed (Münster-Swendsen 1985 and a subsequently updated Turbo Pascal version), including all mortality factors acting in correct sequence and with realistic interactive connections. Fecundity was modeled as a delayed negative feedback function in accordance with the empirical evidence, implicitly containing both sublethal protozoan infections and a dominant unknown factor. Since egg parasitism was generally low and showed no relation to density or weather, it was considered to be a constant estimated by the observed mean value. Mycosis was modeled as a weak delayed negative feedback function using a submodel similar to that for parasitoids (see below). Predation on pupae by nonspecific predators was assumed to be a constant equal to the mean value observed. Parasitism by the two primary parasitoids and one hyperparasitoid was modeled separately, with the effect of each species described by a pseudointerference submodel (Varley et al. 1973, Münster-Swendsen 1985):

$$\log k = \alpha + \beta(\log P), \qquad (2.1)$$

where k is host mortality expressed by the killing power value, P is the density of searching adult parasitoids, and α and β are constants estimated by linear regression of observed k on P. Precipitation was assumed to have a moderate negative effect on needleminer fecundity, thus acting as an exogenous disturbance. The system and conceptual model are shown in figure 2.6.

The simulation model was programmed to allow constants to be changed at will or defaulted to the empirically determined values. Precipitation data were taken from the Danish Meteorological Institute. Simulations with the complete model, empirically derived constants, and observed precipitation produced population cycles with varying amplitudes and 6–7-year periodicity (figure 2.7a), very similar to the observed data (figure 2.1). When precipitation, which acts like a random variable, was set constant, the oscillations dampened with time, approaching a stable equilibrium of 44 needleminer larvae per square meter (figure 2.7b). Moreover, repeated simulations in the presence of identical precipitation data showed that, despite different starting densities, oscillations became synchronous after about 10–20 generations

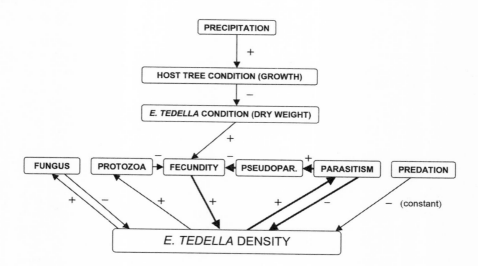

Figure 2.6 Concept for the system model for *E. tedella* population dynamics. In the computer simulation model, the mortalities act in sequence and the fungus disease and predation affect both healthy and parasitized hosts. The less important mortalities due to egg-parasitism and death within the mine are not shown. The negative feedback (loops involving an odd number of minus signs), as indicated for parasitism, pseudoparasitism, sublethal protozoan infections, and fungus disease, all include a delayed response.

The Role of Insect Parasitoids   37

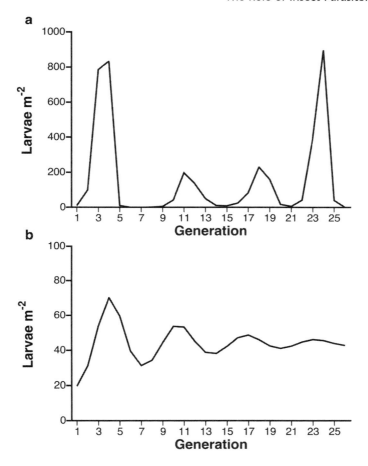

Figure 2.7 (a) Simulation by the complete system model using precipitation data for 25 years. (b) Simulation using constant, average precipitation values.

(years) due to the shared exogenous disturbances (precipitation); that is, the "Moran effect" as described by Royama (1992).

The importance of a given factor in the endogenous feedback structure was evaluated by setting it constant (its feedback was removed), or by setting all the other factors constant (it was the only feedback factor).

(1) When mycosis was kept at its mean value, no detectable change in the simulated population dynamics was observed, and when all other factors were set to their average values while mycosis was the sole feedback factor, the needleminer population stabilized at a very high density (about 2600 larvae/m^2). The inclusion of variable precipitation resulted in extremely violent oscillations with peak densities around 400,000 larvae/m^2. Thus, if mycosis was the only regulating factor, the system would be very unstable and fragile, and complete tree defoliation would frequently occur.

(2) When fecundity was kept at its mean value, the population fluctuations eventually damped to a constant density of 790 larvae/m^2, still much higher than the natural system. With variable precipitation, the needleminer population oscillated continuously with a period of 6–7 years, similar to the natural dynamics, demonstrating that some other factor, such as larval parasitism, was responsible for the cyclic dynamics. However, when fecundity was kept at its maximum value, the population fluctuated around a mean of around 2000 larvae, suggesting that the needleminer population cannot be kept at its observed low density level without fecundity reduction.

(3) Simulations that maintained parasitism and fecundity at their mean levels caused needleminer numbers to reach levels that would completely defoliate the forest. Hence, these two factors, in combination, were necessary to reproduce the observed population fluctuations and mean levels of abundance. Although the inclusion of parasitism alone reproduced the cyclical dynamics, mean host densities were much higher than those observed in nature. Density-induced fecundity reduction was needed to reproduce the observed variations in amplitude and average population densities.

2.4 Time Series Analysis

During the last decade, time series analysis has been widely used by ecologists to diagnose the causes of population fluctuations (e.g., Royama 1977, 1992, Berryman and Millstein 1989, Turchin 1990, Berryman 1992, 1996). In the early 1990s, Alan Berryman analyzed my needleminer and parasitoid time series with his Population Analysis System (Berryman and Millstein 1989, Berryman 1992), and simulations with his simple "logistic" models were compared with those produced by my complex k-factor model (Münster-Swendsen and Berryman in preparation). Correlation analysis *within* the time series (autocorrelation and partial rate correlation functions) indicated that the cyclic dynamics probably resulted from a second-order mechanism, possibly due to a host–parasitoid interaction. Further analysis with a two-species logistic predator–prey model (Berryman 1992) indicated that 74% and 80% of the variation in the per-capita rates of change (R-values) of needleminers and parasitoids, respectively, could be explained by the predator–prey interaction (see chapter 1 and Berryman 1999 for details). This was a surprising result since life table and key factor analyses indicated a much more modest effect of parasitoids, primarily because larval mortality from parasitism only varied between 14% and 78%. In addition, multiple regression analysis of needleminer R-values against percentage parasitism and fecundity reduction indicated that almost all the variation ($r^2 = .76$, $n = 13$) was due to parasitism, and that fecundity reduction did not improve this relationship significantly (multiple $r^2 = .77$, $n = 13$). This was quite surprising, as the key factor (fecundity reduction) was expected to contribute significantly to the determination of R. The question that intrigued us was why time series analysis identified parasitoids as the major contributor to needleminer fluctuations while classical life table analysis identified reduction in fecundity as the key factor? The answer seemed to lie in the discovery of a phenomenon known as pseudoparasitism.

2.4.1 Pseudoparasitism Reduces Fecundity

When parasitoids attempt to deposit eggs in the body of their host, they first inject several biochemical substances, and sometimes a polydnavirus, that weaken the host's immune system and suppress gonadal development (Jones 1985, Brown and Reed 1997). If the host larva escapes, or the parasitoid is disturbed prior to egg deposition, a condition known as pseudoparasitism can occur. Pseudoparasitized hosts survive the attack but are sterile and, because they contain no parasitoid egg or larva, cannot be identified as the victims of parasitism. Could it be that the key factor, reduction in fecundity, is actually caused by parasitoids?

To try to answer this question I first analyzed the relationship between reduction in fecundity (k-values) and successful parasitism (k-values) and found a strong correlation ($r^2 = .796$, $n = 13$, $P < .001$; see figure 2.8). This suggested a close association between these two variables. In addition, the dry weight of unparasitized, hibernating larvae was positively correlated with both succeeding adult fecundity and the degree of larval parasitism (Münster-Swendsen in preparation). I then demonstrated experimentally that pseudoparasitized needleminer larvae are castrated by the event, and finally built a hypothetical model to explain the observed fecundity reductions (Münster-Swendsen 1994). This model predicted the linear relationship between k-values for successful parasitism and k-values for fecundity reduction observed in the data (figure 2.8), and gave a good fit ($r^2 = .78$) to the data when it was assumed that oviposition is interrupted in 75% (3/4) of parasitoid attacks. Figure 2.9 shows the predictions of the model as a function of parasitoid attack frequencies. Notice that when 90% of host larvae are attacked, 50% will contain parasitoid larvae and another 40% will be pseudoparasitized, implying that 80% of the resulting adults will be sterile. This prediction helps to explain why needleminer populations invariably begin to decrease when successful parasitism exceeds 50%, because this actually means that 90% of the population has been rendered impotent by parasitoid attack.

Finally, fecundity reduction was not correlated with larval or adult density and an intensive search for other causes of fecundity reduction, including larval competition, host tree response, and adult dispersal, has proved fruitless. Thus, even though the association between pseudoparasitism and parasitoid density has not been directly measured in the field, it seems fairly obvious that pseudoparasitism (or its share of the fecundity reduction) and successful parasitism should both be directly related to parasitoid density.

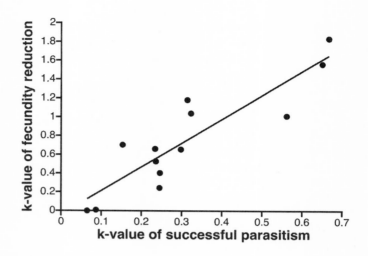

Figure 2.8 Relationship between observed fecundity reduction and successful parasitism in *E. tedella*. Regression equation is $k_F = -0.04 + 2.53(k_P)$, $n = 13$, $r^2 = .796$, $P < .001$.

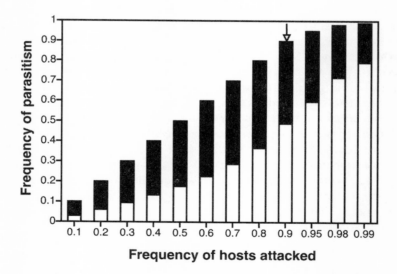

Figure 2.9 Frequencies of successful parasitism (white bars) and pseudoparasitism (black bars) at various frequencies of hosts attacked by the parasitoids. The frequencies of parasitism are predicted by a model based on random search, superparasitism, and a probability of interruption of an attack of 0.75. The arrow indicates the example given in the text.

2.5 Conclusion and Implications

The dynamics of spruce needleminer populations have some special characteristics: (1) the period of oscillation (6–7 years) is relatively contant; (2) peak densities vary considerably but seldom are more than 5% of the spruce needles eaten (well below economic damage levels); and (3) the phase of decrease follows very different peak densities, but always begins when successful parasitism reaches about 50%. Because of (2), and the fact that only old, shaded needles are attacked, the possibility of a significant delayed response by the host tree can probably be excluded. In addition, there is little competition among larvae for appropriate food, even at peak densities.

Time series analysis indicates that the host–parasitoid interaction alone can explain between 70% and 80% of the variation in needleminer and parasitoid population changes from year to year (Berryman 1999). In addition, a model that includes needleminer and parasitoid densities in the preceding generation explains 87% of the variation in needleminer density (Münster-Swendsen in preparation). Parasitism and reduction in host fecundity vary in parallel, are strongly correlated, and together constitute the key factor affecting needleminer fluctuations. The removal of parasitism and density-induced reduction in fecundity from the complex simulation model eliminates needleminer cycles and results in huge needleminer populations capable of completely defoliating the forest. The discovery of pseudoparasitism provides a causative link between parasitoid numbers and the closely related phenomenon of "reduction in fecundity," and explains the conflicting results of key factor and time series analysis. This is an interesting and significant result because it shows how time series analysis identifies the dominant interaction structure of the system better than key factor analysis. The reason for this is that time series analysis and two-species logistic modeling capture the effect of parasitoids on the per-capita rate of change of the needleminer, which includes effects on the *birth rate* as well as the more obvious effects on the death rate, without any knowledge of how the birth rate is affected. Key factor analysis, on the other hand, separates out these effects and, because pseudoparasitism was an unknown entity at the time, was unable to identify the linkage between fecundity and parasitism. For this reason, it underestimates the role of parasitoids in needleminer population fluctuations. From this I conclude that ecologists may benefit from employing time series analysis and logistic modeling before resorting to more complex key factor analysis and detailed lifecycle modeling.

It is difficult to escape from the conclusion that the *specific* cause of population cycles in *E. tedella* is its interaction with insect parasitoids, including the effects of parasitism on the birth rate (pseudoparasitism) as well as the death rate; that is, the structure indicated by heavy arrows in figure 2.6. Sublethal infections by protozoans may amplify the host fecundity effect, since its transmission in this univoltine host species is mainly vertical. The relative constancy of the cycle period facilitates long-term predictions of future peaks (Münster-Swendsen 1987), while logistic models enable 1-year-ahead density predictions that are as good as those of the complex model (Münster-Swendsen and Berryman in preparation). Finally, the knowledge that needleminer populations are almost certain to decrease when more than 50% of the larvae are parasitized, and that variations in peak densities (outbreaks) are determined by random, exogenous disturbances (possibly variations in precipitation) and as yet unexplained, random variations in mortality, are powerful insights for the forest manager.

ACKNOWLEDGMENTS AND COMMENTS

I want to thank Alan Berryman and Judy Myers for their critical and constructive support during the writing of this chapter. In her review, Judy points out that there is no direct proof that parasitoids were responsible for the reduction of host fecundity observed during the study period. In this she is correct, for the level of pseudoparasitism was never measured in the field, nor could it have been since the phenomenon was unheard of at the time of my field studies. Since then, however, the existence of pseudoparasitism in *E. tedella* has been conclusively demonstrated in the laboratory. In addition, the high correlation between adult fecundity reduction and the level of larval parasitism, and between host dry weight, larval parasitism, and adult fecundity reduction, indicates strong functional interrelationships. The total of this "circumstantial" evidence, plus the logical exclusions of other, potentially causative mechanisms, strongly implicates parasitoids as the causal agent of the observed fecundity reduction and the resultant population cycles.

REFERENCES

Berryman, A. A. 1992. On choosing models for describing and analyzing ecological time series. *Ecology* 73: 694–698.

Berryman, A. A. 1996. What causes population cycles of forest Lepidoptera? *Trends Ecol. Evol.* 11: 28–32.

Berryman, A. A. 1999. *Principles of population dynamics and their application.* Stanley Thornes, Cheltenham, UK.

Berryman, A. A. and J. A. Millstein. 1989. *Population analysis system.* Ecological Systems Analysis, Pullman, Wash.

Brown, J. J. and D. A. Reed. 1997. Host embryonic and larval castration as a strategy for the individual castrator and the species. *In* N. E. Beckage (Ed.) *Parasites and pathogens. Effect on host hormones and behavior.* Chapman and Hall, London, pp. 156–178.

Jones, D. 1985. Parasite regulation of host insect metamorphosis: a new form of regulation on pseudoparasitized larvae of *Trichoplusia ni. J. Compar. Physiol. (B)* 155: 583–590.

Münster-Swendsen, M. 1979. The parasitoid complex of *Epinotia tedella* (Cl.) (Lepidoptera: Tortricidae). *Entomol. Meddel.* 47: 63–71.

Münster-Swendsen, M. 1984. The effect of precipitation on radial increment in Norway spruce (*Picea abies* Karst.) and on the dynamics of a lepidopteran pest insect. *J. Appl. Ecol.* 24: 563–571.

Münster-Swendsen, M. 1985. A simulation study of primary, clepto-, and hyperparasitism in *Epinotia tedella* (Cl.) (Lepidoptera: Tortricidae). *J. Anim. Ecol.* 54: 683–695.

Münster-Swendsen, M. 1987. Gode prognoser for grannålevikler-angreb—indtil videre. *Skoven* 19: 96–98.

Münster-Swendsen, M. 1989. Phenology and natural mortalities of the fir needleminer, *Epinotia fraternana* (Hw.) (Lepidoptera, Tortricidae). *Entomol. Meddel.* 57: 111–120.

Münster-Swendsen, M. 1991. The effect of sublethal neogregarine infections in the spruce needleminer, *Epinotia tedella* (Lepidoptera: Tortricidae). *Ecol. Entomol.* 16: 211–219.

Münster-Swendsen, M. 1994. Pseudoparasitism: detection and ecological significance in *Epinotia tedella* (Cl.) (Tortricidae). *Norweg. J. Agric. Sci.* Suppl. 16: 329–335.

Royama, T. 1977. Population persistence and density dependence. *Ecol. Monogr.* 47: 1–35.

Royama, T. 1992. *Analytical population dynamics.* Chapman and Hall, New York.

Turchin, P. 1990. Rarity of density dependence or population regulation with lags? *Nature* 344: 660–663.

Varley, G. C., G. R. Gradwell, and M. P. Hassell. 1973. *Insect population ecology, an analytical approach.* Blackwell Scientific, London.

3

Population Cycles of Small Rodents in Fennoscandia

Ilkka Hanski and Heikki Henttonen

3.1 Introduction

The earliest records of small rodents in Fennoscandia date back to the sixteenth century. Ziegler (1532) and Magnus (1555) reported mass occurrences of the Norwegian lemming (*Lemmus lemmus*), which supposedly descended from the sky, a hypothesis that prevailed for the next 300 years (Henttonen and Kaikusalo 1993)! The first scientific papers on lemmings (Fellman 1848, Ehrstöm 1852) clearly recognized periodicity of lemming dynamics in Finnish Lapland (for a review see Henttonen and Kaikusalo 1993). Collett (1878, 1895, 1911–12) compiled extensive data on lemmings in Norway more than 100 years ago, providing critical material for Elton (1924) to describe the population cycle of small rodents. As these early records suggest, the Norwegian lemming is the most conspicuous member of the small rodent community in northern Fennoscandia, both in appearance and abundance, but apart from mountainous regions, the Fennoscandian small rodent cycle actually refers to *Microtus* and *Clethrionomys* voles rather than to lemmings. At present, the small rodent cycle in Fennoscandia is one of the best documented examples of cyclic population dynamics.

Several recent papers review the state of knowledge on small rodent population dynamics in Fennoscandia and elsewhere (Norrdahl 1995, Krebs 1996, Boonstra et al. 1998, Stenseth 1999, Henttonen and Hanski 2000, Turchin and Hanski 2001). One might think that the "puzzle" of rodent cycles has been solved a long time ago, and that the Fennoscandian small rodent dynamics might serve as a useful reference for the study of cyclic populations in general. Unfortunately, this is not so, although substantial progress has been made over the past 15 years, so that we now have a well-supported hypothesis to explain the small rodent dynamics in Fennoscandia.

There are several reasons why progress has been slow in unraveling the secrets of the small rodent cycle. First, small rodents occur in great abundance throughout the world and there was a tendency to assume that the rodent cycle, especially in northern latitudes, was a universal phenomenon, calling for a universal explanation (Krebs and Myers 1974). However, this is not so. Much of the North American research on small rodents has been conducted on vole populations that are not cyclic in the sense of the definition used in this book (see chapter 1). The North American research tradition has emphasized a syndrome of biological characteristics in "cyclic" voles, such as changes in body size, age at sexual maturity, and demographic rates (Krebs 1996), rather than the usual definition of population cycle in terms of changing numbers of individuals. As a result, researchers have not always agreed on what they should be explaining. The well-documented patterns in vole dynamics in Fennoscandia, summarized in the next section, have occasionally been viewed as a curiosity rather than an important case study on a large spatial scale.

Second, progress has been slow in uncovering the mechanisms of small rodent dynamics because researchers have a fondness for proposing new hypotheses and variants of old ones. According to Batzli (1992), a total of 22 hypotheses have been proposed. Many of these hypotheses have been developed by biologists who tend to focus on biological details, while others have been developed by theoreticians with little knowledge of small rodent biology. Recently, however, opinions seem to be converging on three main plausible hypotheses, albeit still broadly defined, namely predator–prey interaction (rodents being the prey), plant–herbivore interaction, and maternal effects (Batzli 1996, Turchin and Hanski 2001).

Third, it would be foolish to deny the great difficulty of convincingly demonstrating that a particular mechanism drives population cycles in small rodent populations, because such a demonstration would typically require experimental work on an exceptionally large spatial scale. And finally, it is more than likely that different processes may come into play in different small rodent populations in different places and at different times, though the hope is that some generality will eventually emerge.

This chapter first reviews the main patterns in the oscillatory dynamics of small rodents in Fennoscandia. These patterns are well documented and provide an obvious challenge for any hypothesis. Following a description of the patterns, we describe the hypothesis that we consider to be the most plausible explanation for the observed patterns—predation on rodents by small mustelids, weasels and stoats. The following section gives a brief overview of the other hypotheses, which we consider to be less likely explanations of the rodent cycle. The final section attempts to draw some conclusions and to make suggestions for future research. We also outline a new challenge for population ecologists interested in small rodent dynamics, a challenge that goes beyond the study of population cycles.

3.2 Patterns of Oscillatory Dynamics

As this chapter is focused on the dynamics of Fennoscandian small rodents, the observed patterns are described without concern about their similarity to other regions of the world (although this will be discussed in section 3.5).

3.2.1 Pattern 1: Population Cycles Exist

There is no doubt that small rodent populations in Finnish Lapland have cycled, according to the definition used in this book, over long periods of time (figure 3.1), amply demonstrated by the many statistical analyses of these data (Henttonen et al. 1985, Hanski et al. 1993, Turchin 1993, Bjørnstad et al. 1995, Stenseth et al. 1996, Hansen et al. 1999a,b). However, it is equally obvious from figure 3.1 that important changes have occurred in the kind of dynamics and in the species involved, which we have distinguished as a separate pattern.

3.2.2 Pattern 2: Long-term Changes in Relative Abundance and Type of Dynamics

At Kilpisjärvi in Finnish Lapland, *Clethrionomys rufocanus* was the dominant vole species from the 1950s until the end of the 1980s, with the exception of peaks by the Norwegian lemming in 1960 and 1969–70 (figure 3.1a). During the past 10 years, however, the pooled density of other species, primarily *Clethrionomys rutilus*, has typically far exceeded that of *C. rufocanus* (figure 3.1a). At Pallasjärvi, 200 km southeast from Kilpisjärvi (figure 3.2), the dynamics of all coexisting species were cyclic in the 1970s (Henttonen 1985, Henttonen et al. 1987) but, since the early 1980s, the density of *Microtus agrestis* has remained low while annual fluctuations have been dominated by *Clethrionomys glareolus* (figure 3.1b). During the latter period, time series analysis indicates no significant delayed density dependence in the dynamics of *C. glareolus* (Yoccoz et al. 2000). These examples illustrate that much more is involved in the Fennoscandian small rodent communities than "just" a regular population cycle.

3.2.3 Pattern 3: Latitudinal Gradients

There is a definite geographical pattern in both the amplitude and period of the cycle, which both increase with latitude in Fennoscandia (Hansson and Henttonen 1985a,b, 1988, Henttonen et al. 1985, Henttonen and Hansson 1986, Hanski et al. 1991, 1993, Turchin and Hanski 1997). For example, the period is around 5 years at Kilpisjärvi (figure 3.1a) but only 3 years in southern Finland (figure 3.2). At latitudes lower than around 60°N, regular multiannual periodicity disappears and only the annual fluctuations remain (Hansson and Henttonen 1985a, Hanski et al. 1991). Cycle amplitude, as measured by the ratio of the maximum to the minimum density, is more

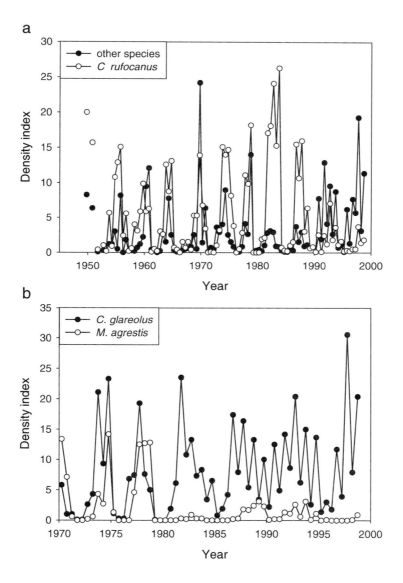

Figure 3.1 Long-term record of small rodent populations in two localities in Finnish Lapland, Kilpisjärvi (a) and Pallasjärvi (b); for their locations, see figure 3.2. In (a), numbers are shown separately for *C. rufocanus* and the other species added together, including *L. lemmus* (very abundant in the early 1970s) and *C. rutilus*. In (b), numbers are shown separately for *M. agrestis* and *C. glareolus* (for the other species see Henttonen et al. 1987, Hanski and Henttonen 1996).

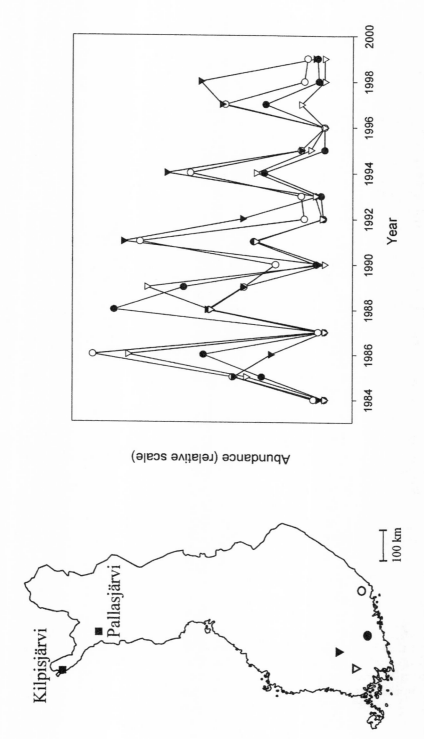

Figure 3.2 Changes in the numbers of ringed fledglings of the Tengmalm's owl at four localities in southern Finland, used as a surrogate for the size of the local vole populations (the owl breeding effort and success are closely related to the availability of small rodent prey). The map shows the locations of the four sites as well as the two well-studied localities in Lapland, Kilpisjärvi and Pallasjärvi.

than 100 in the north and only around 10 in southern Fennoscandia (Hanski et al. 1991).

3.2.4 Pattern 4: Geographical Synchrony

Populations typically show a high level of geographical synchrony over large areas (figure 3.2; for Lapland see Heikkilä et al. 1994, fig. 2; see also Henttonen et al. 1977, Andersson and Jonasson 1986, Henttonen and Wallgren 2000), though clearly not over the whole of northern Fennoscandia due to the latitudinal gradient in average cycle period (Pattern 3 above).

3.2.5 Pattern 5: Interspecific Synchrony

Along with increasing amplitude and cycle period, synchrony between different rodent species also increases with latitude (Henttonen and Hansson 1986, Hanski 1987). Interspecific synchrony is particularly apparent in northern populations (figure 3.1). An exception is the latter part of the record from Pallasjärvi (figure 3.1b), but this exception tends to prove the rule that interspecific synchrony occurs whenever high-amplitude oscillations prevail. In particular, high-amplitude oscillations are usually associated with synchronous population declines that typically extend into summer (Hansson and Henttonen 1985a, 1988). It is also worth noting that cycle amplitude is typically greater in *Microtus* voles than in *Clethrionomys* voles (Hanski and Henttonen 1996), which could be distinguished as yet another important pattern.

3.2.6 Pattern 6: The Norwegian Lemming

The dynamics of *L. lemmus* differ drastically between south-central and northern Fennoscandia. In the high mountain area of southern Norway, lemmings exhibit regular 3–4-year cycles (Framstad et al. 1997). In Lapland, however, the pattern is much more irregular, with some lemming peaks coinciding with vole peaks and others not (figure 3.1a). Either lemming populations totally miss some vole peaks, or the increase is too small to observe. In alpine areas, lemmings may cycle for extended periods with such a low amplitude that the "cycle" is not observed without special trapping (Virtanen et al. 1997). In Lapland, lemming peaks occur more often in the northern mountainous areas than further south in the taiga forest (Henttonen and Kaikusalo 1993). These authors suggested that the regularity of the lemming cycle is related to the availability of extensive snow-covered areas that are important for winter breeding. Snow depth and coverage are much greater in the high mountains in southern Norway than in northern Fennoscandia.

The great lemming outbreaks, with their spectacular long-distance migrations that may extend south of the Arctic Circle in Finnish Lapland, exhibit a persistent long-term pattern (Henttonen and Kaikusalo 1993). Great lemming years have occurred in 1755, 1787, 1810, 1840, 1872, 1902–3, 1938, and 1970, at roughly 30-year intervals. Following their movement into the taiga forest, lemmings may survive there for some time. For example, following the 1970 migration, local lemming peaks were again observed in many areas in Lapland in synchrony with the vole peaks in 1974, 1978, and 1982. During outbreak years, lemmings are the most abundant small rodents in all habitats of the northern taiga as, for instance, at Pallasjärvi in 1970 (Henttonen et al. 1987, Henttonen and Kaikusalo 1993). At such times they have a strong impact on other microtine species and vegetation (Kalela and Koponen 1971, Oksanen and Oksanen 1992).

3.2.7 Pattern 7: The Chitty Effect

Boonstra and Krebs (1979) coined the term "Chitty effect" to designate a "syndrome of characteristics" of population cycles. This was later called the "biological definition" of cyclic fluctuations (Krebs 1996). Although we use the familiar numerical definition of population cycles in this book, it is worth adding the Chitty effect here as yet another pattern related to population cycles of small rodents. Chitty (1952, 1960) originally defined the periodic fluctuation of vole populations by their associated demographic and physiological characteristics, the most prominent of which are phase-related changes in body size; that is, voles are large in the increase and peak phase and small in the decrease and low phase. Phase-related changes in body size occur in Fennoscandia (Sundell and Norrdahl in preparation, Henttonen unpublished), and the latitudinal gradient (Pattern 3) is correlated with variation in body size and life history traits (Hansson and Henttonen 1985b).

3.3 The Predator–Prey Hypothesis

Beginning with the pioneering study of Anderson and Erlinge (1977), research on small rodent dynamics in Fennoscandia during the past 20 years has, to a large extent, focused on predation as the key ecological interaction maintaining and adjusting the rodent cycle. The predation hypothesis as applied to small rodents in Fennoscandia has two components, involving specialist predators and generalist predators, respectively.

3.3.1 Specialist Predators

There are two resident specialist predators in Fennoscandia, the least weasel (*Mustela nivalis*) and the stoat (*Mustela erminea*). These predators respond numerically to changes in the density of their main prey, as they have little in the way of alternative food (Korpimäki et al. 1991, Korpimäki 1993) and have limited ability to move elsewhere when prey becomes scarce. The main prey of the least weasel is the field vole, *M. agrestis*, while stoats in Lapland prey mainly on the large root vole, *Microtus oeconomus*. In southern Sweden and Finland, the stoat has a more diverse diet (Erlinge 1975, Korpimäki et al. 1991), and acts both as a resident specialist and a resident generalist depending on the prey species present in the local community. In developing the predator–prey model (below), we have the least weasel and field vole in mind, and the model parameters have been estimated for these two species (Hanski et al. 1993, Hanski and Korpimäki 1995, Turchin and Hanski 1997).

Resident specialist predators with a delayed numerical response are considered to drive the population oscillations of small rodents in northern Fennoscandia (Hansson 1987, Henttonen 1987, Henttonen et al. 1987, Hanski et al. 1991, 1993, 1994, Korpimäki et al. 1991, Heikkilä et al. 1994, Hanski and Korpimäki 1995). It is a well-established theoretical tenet that interactions between specialist predators and their prey can generate and maintain population oscillations (e.g., May 1991). We have developed the following model to describe the numerical interaction between small rodents and mustelids:

$$\frac{dN}{dt} = aN\left(1 - \frac{N}{K}\right) - \frac{cNP}{D+N},$$
$$\frac{dP}{dt} = vP\left(1 - \frac{qP}{N}\right),$$

(3.1)

where N and P are the prey and the predator population sizes, respectively. The model assumes logistic growth for the prey population in the absence of predation, a type II functional response for the predator, and logistic growth for the predator population with "carrying capacity" proportional to the current prey population size (Leslie 1948, May 1973, Tanner 1975). This model has an identical structure, although a somewhat different interpretation, to the logistic "ratio-dependent" model developed by Berryman (1992, 1999b). For more details see Hanski et al. (1991, 1993, 2001), Hanski and Korpimäki (1995), and Turchin and Hanski (1997), who also describe how seasonality was added to the model. One version of the model (Hanski et al. 1993) also includes a threshold prey density for predator reproduction, a phenomenon which has been observed empirically (Erlinge 1974) and which eliminates the possibility of predator reproduction when both N and P/N are small. Note that the model includes direct density dependence in rodent dynamics due to intraspecific competition for space or food resources, hence the model does not imply that predation alone is responsible for the predicted dynamics.

The predator–prey model with seasonality is a three-dimensional continuous-time model and may consequently exhibit a range of dynamics from stable equilibrium to chaos. However, even in the chaotic regime, the oscillations involve a distinctly regular component (figure 3.3), reflecting the intrinsic tendency to cycle in the two-dimensional predator–prey model. Using parameter values estimated from independent data, as explained in Hanski et al. (1993), Hanski and Korpimäki (1995), and Turchin and Hanski (1997), the model predicts dynamics that closely resemble those

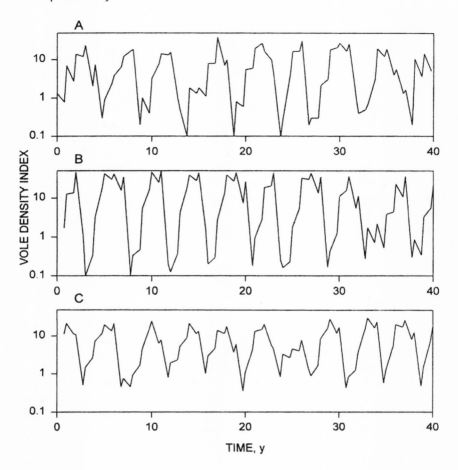

Figure 3.3 Comparison between the observed dynamics at Kilpisjärvi in 1952–92 (A, two data points per year) and the predicted dynamics by two versions of the predator–prey model (B, Turchin and Hanski 1997; and C, Hanski et al. 1993). From Turchin and Hanski 1997.

observed, in terms of amplitude, average period, and strength of the periodicity as measured by the autocorrelation function (figure 3.3; Hanski et al. 1993, Hanski and Korpimäki 1995, Turchin and Hanski 1997). Turchin and Ellner (2000) demonstrated that the model predicts more than half the variance in the data from Kilpisjärvi (prediction $r^2 = .59$, where the future value N_{t+1} is predicted from the present values N_t and P_t). These results provide quantitative support for the predator–prey hypothesis explaining Pattern 1.

Some ecologists (e.g., Ginzburg 1998) prefer an alternative equation for the predator population, in which the numerical response is obtained by multiplying the prey killed by the functional response and by a reproductive constant, what Ginzburg calls the "conversion principle," but more generally known as the Rosenzweig–MacArthur (1963) model. The reason we used a logistic predator model is discussed by Hanski et al. (2001). In addition, Berryman (1999a) has argued that logistic survival models, like (3.1), are more appropriate for describing the dynamics of populations made up of individual organisms than are those based on the conversion principle. Turchin and Ellner (2000) used both types of models to describe the time series from Kilpisjärvi (figure 3.1a) and found that the Rosenzweig–MacArthur model fitted the data equally well but, in so doing, yielded unrealistic values for one of the parameters. This result provides support for the model structure we employed. Notwithstanding this difference of opinion about model structure, the main conclusion is that both types of simple predator–prey models predict the same dynamic pattern as observed in the data.

3.3.2 Generalist Predators

Apart from small mustelids, small rodent populations in Fennoscandia are consumed by a diverse array of generalist mammalian and nomadic avian predators (Henttonen and Hanski 2000). Resident generalists, such as the red fox, badger, and feral cats, may switch to alternative prey when rodents become scarce (Erlinge et al. 1983). Consequently, these predators show a weaker numerical response to rodent numbers than do resident specialist predators, and may therefore have a type III or similar functional response (Turchin and Hanski 1997). In the north, however, where there are fewer alternative prey, species like the red fox become more closely linked to rodent dynamics. Nomadic avian predators are represented by several species of owls and raptors (Korpimäki and Norrdahl 1991a, Korpimäki 1993). Although these species are rodent specialists, their mobility enables them to track changing density of prey in space (Andersson and Erlinge 1977, Löfgren et al. 1986, Korpimäki et al. 1987) and, therefore, they function much like resident generalists.

Empirical data indicate that the pooled density of nomadic specialists and generalist predators increases by an order of magnitude from the north ($70°C$) to the south ($56°C$) in Fennoscandia (Hanski et al. 1991). These authors also demonstrated that increasing predation pressure by such predators increases the stability of rodent populations, decreasing both the amplitude and periodicity of the oscillations. Turchin and Hanski (1997) went on to include generalist predators in the model, with parameters estimated from independent data. This model predicted, in quantitative terms, the increasing oscillatory tendency of rodent fluctuations observed with increasing latitude (Pattern 3). The predicted patterns and their ACFs agreed remarkably well with the observed patterns at different latitudes (figure 3.4). Our conclusion is that the predation hypothesis provides a plausible quantitative explanation for Pattern 3.

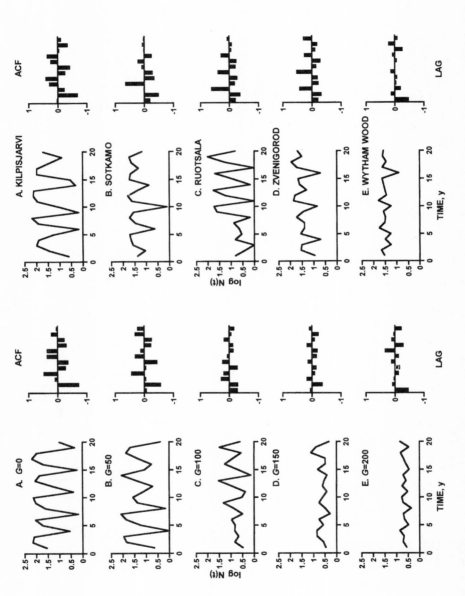

Figure 3.4 Left columns: The rodent dynamics predicted by the model of Turchin and Hanski (1997) for different densities of generalist predators (G). Right columns: The observed dynamics at different latitudes from north to south (Kilpisjärvi 69°N, Sotkamo 64°N, Ruotsala 63°N, Zvenigorod 57°N, and Wytham Wood 51°N). From Turchin and Hanski 1997.

3.3.3 The Other Patterns

The two components of the specialist/generalist predation hypothesis explain Patterns 1 and 3 in the dynamics of small rodent populations in Fennoscandia. According to this hypothesis, resident specialists (small mustelids) are instrumental in maintaining high-amplitude oscillations (Pattern 1), while latitudinal gradients in the density of generalist predators and nomadic avian predators generate increasing stability (lower amplitude and period) as one moves south (Pattern 3). We will address Pattern 2 in section 3.5, Pattern 7 in section 3.4, and the three remaining patterns below.

The large-scale geographical synchrony of rodent oscillations (Pattern 4) has been explained by the effects of nomadic avian predators (Ydenberg 1987, Korpimäki and Norrdahl 1989, Ims and Steen 1990). The idea is simple: If a local population is out of synchrony, nomadic avian predators, with their ability to quickly assess local rodent densities in spring, settle in large numbers to breed near abundant prey enclaves and force the populations back to the regional cycle. Heikkilä et al. (1994) suggest that stoats may play a similar role in Lapland due to their relatively high mobility.

A different explanation for regional synchrony, one that does not involve predators, is the so-called Moran effect (Moran 1953, Royama 1992). This explanation assumes that local populations are all under the control of the same density-dependent mechanism and that, if all populations are "set" to a similar density on occasion, they would then oscillate in synchrony for a time simply because they all obey the same laws of population change. As such, any large-scale major environmental disturbance might suffice to synchronize populations. In an example described by Sheftel (1989), an early snow melt is followed by very cold weather in spring, causing a premature decline of cyclic shrew populations in central Siberia.

The small rodent–predator community in Fennoscandia contains a large number of species (Hanski et al. 1991, Henttonen and Hanski 2000). Although the predation hypothesis focuses primarily on just two species, the field vole and the least weasel, multiannual oscillations of other small rodents could be driven by the interaction between these two common and widespread species via competitive interactions and predator switching. Thus Pattern 5, interspecific synchrony, could be explained by the effects of shared predators. Interspecific synchrony includes insectivorous shrews, to the extent that population declines of shrews tend to be synchronous with population declines of voles (Hansson 1985, Henttonen 1985, Kaikusalo and Hanski 1985, Henttonen et al. 1989). Such synchronous declines of species with entirely different biology (shrews vs. voles) are difficult to explain by factors other than predation. Support for the predation hypothesis is also provided by direct observations of radio-collared voles, in which most of the mortality during vole declines has been shown to be caused by mustelids (Norrdahl and Korpimäki 1995b, Steen 1995). Predation, therefore, provides a logical and empirically supported explanation for Pattern 5.

The dynamics of the Norwegian lemmings have been explained both by food supply and predation, but this dichotomy refers to different habitats with dissimilar productivities (Kalela and Koponen 1971, Oksanen and Oksanen 1992, Henttonen and Kaikusalo 1993, Turchin et al. 2000). In alpine (low-productive) habitats, lemmings are apparently regulated by food supplies. In this case lemmings act as the "predators" and plant food the prey. In the more productive habitats of the birch forest zone and taiga (following large-scale lemming migrations), lemmings seem to be regulated by the same predator community that causes vole oscillations (lemmings now act as "prey"). In the Pallasjärvi taiga, lemming declines were synchronous with those of voles (Henttonen et al. 1987). In addition, the disappearance of lemmings from the taiga after periodic outbreaks has usually been explained by predation (Henttonen and Kaikusalo 1993), because the large and clumsy lemmings are much more susceptible to predators than the smaller and more agile voles. During low densities in the birch forests of Kilpisjärvi during 1998, radio-collared lemmings were mostly killed by predators (Gower and Henttonen unpublished). It is common to find fresh lemming carcasses and, when carefully examined, the cause of death is usually predation, especially by owls (Steen 1995 and our own observations).

3.4 Other Hypotheses

Chapter 1 describes six principal hypotheses concerning the causes of cyclic population dynamics in general. In this section we briefly discuss the relevance of the remaining five hypotheses, apart from predation, to the rodent cycle in Fennoscandia.

3.4.1 Physical Effects

Temporal variation in sunspot activity is much longer than 5 years, the longest period observed in small rodent oscillations in Fennoscandia. No periodic climatic factor of sufficient strength, large enough amplitude, and appropriate period has been documented.

3.4.2 Parasites and Pathogens

Small rodents have a large number of macroparasites (such as tapeworms) and microparasites (such as viruses), many of which are specific to individual species or groups of rodent species (Gonzales and Duplantier 1999, Plyusnin et al. 1999, Haukisalmi and Henttonen 2000). It is evident that high-amplitude oscillations of the host species have a large impact on the dynamics of the parasites (Brummer-Korvenkontio et al. 1982, Keith et al. 1985). It is equally clear that particular pathogens have caused high mortality in particular small rodent populations at particular times (Soveri et al. 2000). In the present context, however, the crucial question is whether parasites have a strong and consistent enough effect to generate population cycles of the correct period and amplitude. The search for evidence of parasite-driven rodent cycles has continued since Elton's pioneering research (Elton et al. 1935), but no well-documented case study yet exists. We may ask why that should be so.

Macroparasites typically do not directly increase the mortality of their rodent host. Furthermore, a 3–5-year cycle is too short for macroparasites, with their complicated life cycles, to track the rodent cycle (Haukisalmi and Henttonen 1990). However, similar macroparasites can track the 10-year snowshoe hare cycle (Keith et al. 1985). Diseases caused by mammalian microparasites may interact with many factors (e.g., the functioning of the immune system depends on age, sexual status, and nutrition), but these interactions are probably complex and unlikely to lead to simple population dynamic patterns. Soveri et al. (2000) suggest that diseases may modify rodent dynamics generated mainly by predator–prey interactions.

3.4.3 Plant–Herbivore Interaction

This hypothesis has received substantial attention among ecologists working with small rodents since the papers by Lack (1954) and Pitelka (1958), as well as the pioneering research by Kalela (1962) in Fennoscandia. For reviews see Batzli (1983) and Turchin and Batzli (2001). The basic idea is that, in one sense or another, rodents overexploit their food resources, they do not reproduce well unless high-quality food is abundant, and inadequate food triggers population declines. There are indeed multiannual oscillations in plant quantity and quality in northern Fennoscandia, to some degree correlated with the changing density of small rodents (Tast and Kalela 1971, Laine and Henttonen 1983). However, more recent research suggests that plant quality has little impact on rodent reproduction and survival, so that the feedback from plants to rodents must be weak or nonexistent (Jonasson et al. 1986, Laine and Henttonen 1987).

A recent experiment provides convincing evidence against the plant–herbivore hypothesis. Klemola et al. (2000) established populations of field voles in four large predator-proof enclosures that had been heavily grazed by voles the preceding autumn and winter. Control enclosures had no such history of previous grazing. These authors found no detrimental effects of previous grazing on the reproduction or body condition of voles, nor on the population growth rate (Klemola et al. 2000). A similar experiment in North America reached the same conclusion (Ostfeld et al. 1993). These studies strongly suggest that oscillations of *Microtus* in grassy habitats are not driven by delayed effects of the plant–herbivore interaction.

3.4.4 Maternal Effects

High density and food shortages can obviously have adverse effects on reproducing females that may influence the performance of their offpsring, and such effects could generate time delays and consequent cycles in rodent dynamics. Inchausti and Ginzburg (1998) have constructed a mathematical model of this hypothesis, which is capable of generating 3–5-year cycles. However, the longer periods seen in many small rodent populations require unrealistically small values for the intrinsic rate of increase (r) of the animals (i.e., $r = 2.46$ in their fig. 5c, whereas realistic values are in the range 5 to 7, Turchin and Hanski 1997, Turchin and Ostfeld 1997). The claim that the maternal effect explains the latitudinal gradient in cycle period (Pattern 3) is based on the assumption that the realized per-capita rate of change decreases with increasing latitude, but there is no evidence of such an effect (Hanski et al. 1991). Furthermore, the model of Inchausti and Ginzburg (1998) does not explain the absence of multiannual cycles in southern Fennoscandia. Thus, while maternal effects can generate, in principle, the kind of delayed density dependence necessary for multiannual population cycles, the lack of empirical evidence for a strong maternal effect, together with unconvincing modeling exercises, gives little support to this hypothesis.

3.4.5 Genetic Effects

The principal idea here is that density could impose a sufficiently strong selection pressure to lead to rapid changes in the genetic composition of the population, which in turn would influence the rate of change of population size (Chitty 1967). This idea, though extremely influential amongst ecologists working on small rodents (especially in North America) during the 1970s and early 1980s, has received little empirical support and is widely considered to be refuted (Stenseth 1999). Intriguingly, a recent study by Sundell and Norrdahl (in preparation) suggests that the Chitty effect may be partly due to selective predation on larger voles during the decline and low phases. These authors demonstrate that voles smaller than 20 g can enter holes too small for even the smallest weasels and, therefore, have a refuge from predation.

In Fennoscandia, latitudinal changes in rodent body size and life history traits have been noted by Hansson and Henttonen (1985b). In our view, this variation is likely to reflect adaptive responses of voles to environmental variation, possibly including differences in the type of dynamics. We do not deny that these differences could have some population dynamic consequences, but there has been no clear argument, supported by quantitative models, showing how these differences in body size and life history traits could cause population cycles in the north but not in the south (Patterns 1 and 3).

Another intrinsic mechanism that has been discussed in the context of small rodent cycles is the "stress hypothesis" (Christian 1978), but current thinking is that the effect of stress on behavioral–endocrine systems is unlikely to be the main cause of population cycles in small rodents. However, expanding knowledge on the interaction between stress, nutrition, and immunological function may produce interesting insights, including the effects of stress caused by the threat of predation.

Our overall conclusion is that no substantial body of evidence supports any of the other hypotheses for cycles in small rodent populations in Fennoscandia. In particular, recent experimental evidence strongly argues against the plant–herbivore hypothesis. Furthermore, all the alternative hypotheses have only been employed to explain Pattern 1 (with the exception of the maternal effect hypothesis and Pattern 3, see above). It is difficult to see how these mechanisms might explain, even in principle, most of the other patterns described in section 3.2.

3.5 Discussion

The evidence that interactions with predators are responsible for small rodent oscillations in Fennoscandia, explaining Patterns 1, 3, 4, and 5 (section 3.2), is quite strong but not yet overwhelming. As discussed in section 3.3, predation could even explain Pattern 7, the "Chitty effect," which many North American ecologists have previously interpreted as prime evidence for some "intrinsic" mechanism. The modeling work (section 3.3, Hanski et al. 2001) clearly demonstrates that predator–prey interactions provide a plausible quantitative explanation for the Fennoscandian rodent cycle, but cannot exclude the possibility that, someday, an equally satisfactory explanation will involve another mechanism. Field observations are generally consistent with the predation hypothesis (Korpimäki and Norrdahl 1991a,b, Korpimäki 1993, Norrdahl and Korpimäki 1995b, Steen 1995, and many others, including the studies reported in section 3.2), but these alone can never be conclusive. The few experimental studies that have been attempted (Norrdahl and Korpimäki 1995a, Korpimäki and Norrdahl 1998; for a review see Hanski et al. 2001) are also consistent with, or at least do not contradict, the predation hypothesis. Unfortunately, the logistical problem of conducting large-scale experiments on voles and weasels remains a major obstacle. It is difficult to see at this point how a single critical experiment could be designed to resolve the long-standing "puzzle" once and for all. Rather, we expect to rely on accumulating field observations, theoretical modeling, and field experiments, which currently support the predation hypothesis.

Although this chapter is primarily concerned with small rodent dynamics in Fennoscandia, it is helpful to add some comments about the applicability of our ideas to other regions of the world. Cyclic rodent populations, in many ways similar to those in Fennoscandia, have been studied in Hokkaido, Japan (Saitoh 1987, Stenseth and Saitoh 1998) and in northern England (Lambin et al. 2000). Even in the most cyclic region on Hokkaido, only one third of the populations show significant periodicity, and even these seem to behave more like the populations at the transition zone (around 60°N) in Fennoscandia (Hansson and Henttonen 1998). Unfortunately, very little information is available on rodent predators, apart from the fact that Hokkaidon predator communities include many generalists (Henttonen et al. 1992).

In northern England, where Lambin and his colleagues have conducted intensive studies on the field vole and its predators, a distinct vole cycle with a period of 3–4 years was apparent in the 1980s and 1990s (Lambin et al. 2000, fig. 3). In a quantitative sense, Lambin's results do not support the model of Turchin and Hanski (1997), because the pressure by generalist predators is so high that the model, with parameters estimated from Fennoscandian data, predicts stable dynamics. However, field vole oscillations in northern England have very small amplitude (ratio of maximum over minimum densities around 5), comparable with those in southern Fennoscandia. The surprising thing is the exceptional regularity of the cycles in northern England (five cycles documented in detail). The cause of these cycles may still be predation by weasels, which occur in high densities (Lambin et al. 2000), and experiments are currently under way to test this hypothesis (Lambin personal communication).

The data from boreal North America indicate a more stable and less cyclic small rodent community than in Fennoscandia (Hansson and Henttonen 1985a, Henttonen et al. 1985, Taitt and Krebs 1985). However, the contrast is less striking when one realizes that these studies have focused on *Clethrionomys* species rather than on *Microtus* (e.g., Taitt and Krebs 1985). Based on the Fennoscandian results (figure 3.1b), we would not expect regular, high-amplitude cycles in small mammal communities dominated by *Clethrionomys*, with the exception of the large, *Microtus*-like *C. rufocanus* (figure 3.1a). The predation hypothesis predicts that Fennoscandian-like small mammal dynamics will be observed in boreal North American communities with extensive areas of high-quality, grassy habitat suitable for *Microtus*. This prediction appears to be supported by results from central Alaska, where *Clethrionomys* are relatively stable but *Microtus* cycle (Whitney and Feist 1984). Results from Siberia are also consistent with this prediction. At a study site on the floodplains of the river Yenisei, with a diverse community of small mammals in very productive habitats, a distinct 4-year small mammal cycle has been documented (Sheftel 1989). In contrast, many other study sites in Siberia are dominated by *Clethrionomys*, most likely because of less productive habitats, and these populations are not cyclic in the Fennoscandian sense (Henttonen et al. 1985).

We finally return to Pattern 2, the long-term changes that are observed in the relative abundance and kind of dynamics exhibited by some small rodent populations in Fennoscandia. The important question for us is whether these observations refute the predation hypothesis, and more generally, what are the implications for the study of population cycles?

There has undoubtedly been a substantial change in the type of dynamics exhibited by *M. agrestis* and *C. glareolus* at Pallasjärvi in the mid-1980s, in the sense that cycles appear to be disappearing from this region (Henttonen et al. 1987, figure 3.1b), as well as elsewhere in northern Fennoscandia (Hörnfeldt 1994, Lindström and Hörnfeldt 1994, Hansson 1999, Henttonen and Wallgren 2000). A more general issue concerns dramatic changes in the relative abundance of coexisting species that seem to be associated with these changes in small rodent dynamics. Several explanations have been suggested. First, the changes may have been caused by a large-scale environmental change as, for instance, in the structure of forested landscapes caused by forestry practices (Hansson 1999) or climate change. However, there is no actual evidence to support such an argument. Indeed, Lambin et al. (2000) make the point that field voles in northern England have been cyclic for the past 60 years in spite of great changes in the structure of the forested landscape. (This does not necessarily refute Hansson's 1999 idea, however, because he based it on the availability of lichens in the boreal old-growth forests, while English forests are manmade plantations.)

Within the framework of the predation hypothesis, the reduced amplitude of population cycles in the north can be explained by increasing numbers of generalist predators, which are expected to have a stabilizing influence (section 3.3). The red fox and the American mink are possible candidates. However, this argument does not readily explain why a previously dominant species, *M. agrestis* at Pallasjärvi (figure 3.1b) and *C. rufocanus* at Kilpisjärvi (figure 3.1a), has failed to reach previous peak densities.

Another possible explanation for Pattern 2 relates to interspecific interactions within multispecies communities in Fennoscandia (Henttonen 1987, 2000, Henttonen et al. 1987, Hanski and Henttonen 1996, Henttonen and Hanski 2000). Given the strong interactions typical of many vertebrate predator–prey communities, it is possible that at least some aspects of the dynamics are determined by complex multispecies interactions, including indirect interactions. The issue goes to the heart of current concern about the future of community ecology. Several prominent ecologists (e.g., Brown 1999, Lawton 1999) have recently doubted whether predictive community ecology will ever be possible, because of the dynamic complexity that strong multispecies interactions generate (incidentally, these apprehensions have been expressed in support of macroecological studies). Microcosm experiments provide examples where species composition diverges widely in similar replicate communities (Niederlehner and Cairns 1994, Balciunas and Lawler 1995, McGrady and Morin 1996), probably because of complex interspecific interactions. In the case of small rodents, it is possible that long-term changes in species composition and population dynamics are due to complex interactions in multispecies communities, possibly triggered by some (minor) environmental changes. If so, we believe that the challenge should be accepted and a mechanistic explanation of these changes should be sought, however difficult that might be.

Hanski and Henttonen (1996) have made a start by extending the two-species model to three species—least weasel, field vole, and bank vole. The main empirically derived assumption is that the saturation parameter [D in equation (3.1)] of the weasel functional response to the bank vole is larger than that to the field vole, making the interaction with the bank vole more stable (less efficient predator). In addition, it is known that the larger field vole is a superior competitor to the bank vole (Grant 1972, Hansson 1983). With these two assumptions and realistic parameter values, the model predicts the two major patterns in multispecies rodent oscillations. First, during one population cycle, the abundance of the bank vole is predicted to peak earlier than that of the field vole, because weasels suppressed the abundance of the latter to a lower level during the previous low phase (because of the smaller D value), and because the field vole competitively suppresses the bank vole during the peak. Empirical results support this prediction (Hanski and Henttonen 1996, figs. 2 and 3). This mechanism could also explain why population oscillations have a greater amplitude in *Microtus* than in *Clethrionomys* (part of Pattern 5).

Second, for a range of parameter values and with some environmental stochasticity added to the model, the predicted dynamics show long periods of numerical dominance by one of the vole species, followed by a sudden switch of their relative abundances. Furthermore, periods of high-amplitude oscillation are associated with dominance by the field vole. These predictions have the following intuitive explanation: During periods of field vole dominance, the interaction between this species and weasels dominates the three-species dynamics, forcing synchronous oscillations in the bank vole because of shared predation. In contrast, during periods of bank vole dominance, the more stable interaction between this species and the weasel (large D) becomes dominant, largely eliminating the multiannual oscillatory tendency. The field vole population is kept at a low level because of high and relatively constant predation, maintained by the bank vole population. This is a plausible explanation for the switch in dynamics evident in the long time series for Pallasjärvi (figure 3.1b). An additional factor may be an apparent shift in numerical dominance among small mustelids, from the least weasel to the stoat in Lapland (Henttonen et al. 1987, Oksanen et al. 1999, Henttonen 2000).

The message here is that complex interactions in multispecies communities may generate complex dynamics with sudden changes in pattern that extend over long periods of time. Given that the Fennoscandian small rodent–predator communities contain 15–20 species, often with strong interspecific interactions, such dynamic complexity is not surprising. This may lead one to conclude that the cynical views of Brown (1999) and Lawton (1999) about predictive community ecology are justified. Nevertheless, we emphasize that it is possible to explain the patterns described in section 3.2 within the general framework of the predation hypothesis. From this standpoint, we agree with the ideas outlined in chapter 1 concerning the general significance of food web architecture in population oscillations (Pimm 1991), of which the small rodent dynamics in Fennoscandia provide a prime example (Hansson and Henttonen 1988).

ACKNOWLEDGMENTS

We would like to acknowledge the helpful comments and suggestions of Xavier Lambin, Mauricio Lima, and Alan Berryman on earlier versions of this chapter.

REFERENCES

Anderson, M. and M. Erlinge. 1977. Influence of predation on rodent populations. *Oikos* 29: 591–597.
Andersson, M. and S. Jonasson. 1986. Rodent cycles in relation to food resources on an alpine heath. *Oikos* 46: 93–106.
Balciunas, D. and S. P. Lawler. 1995. Effects of basal resources, predation, and alternative prey in microcosm food chains. *Ecology* 76: 1327–1336.
Batzli, G. O. 1983. Responses of arctic rodent populations to nutritional factors. *Oikos* 46: 93–106.
Batzli, G. O. 1992. Dynamics of small mammal populations: a review. *In* D. R. McCullough and R. H. Barret (Eds.) *Wildlife 2001: Populations*. Elsevier, London, pp. 831–850.
Batzli, G. O. 1996. Population cycles revisited. *Trends Ecol. Evol.* 11: 488–489.
Berryman, A. A. 1992. The origin and evolution of predator–prey theory. *Ecology* 73: 1530–1535.
Berryman, A. A. 1999a. Alternative perspectives on consumer–resource dynamics: a reply to Ginzburg. *J. Anim. Ecol.* 68: 1263–1266.
Berryman, A. A. 1999b. *Principles of population dynamics and their application*. Stanley Thornes, Cheltenham, UK.
Bjørnstadt, O. N., W. Falck, and N. C. Stenseth. 1995. A geographic gradient in small rodent density fluctuations: a statistical model approach. *Proc. Roy. Soc. Lond. Ser. B* 262: 127–133.
Boonstra, R. and C. J. Krebs. 1979. Viability of large- and small-sized adults in fluctuating vole populations. *Ecology* 60: 567–573.
Boonstra, R., C. J. Krebs, and N. C. Stenseth. 1998. Population cycles in small mammals: the problem of explaining the low phase. *Ecology* 79: 1479–1488.
Brown, J. H. 1999. Macroecology: progress and prospects. *Oikos* 87: 3–14.
Brummer-Korvenkontio, M., H. Henttonen, and A. Vaheri. 1982. Hemorrhagic fever with renal syndrome in Finland: ecology and virology of *Nephropathia epidemica*. *Scand. J. Infect. Dis.* Suppl. 36: 88–91.
Chitty, D. 1952. Mortality among voles (*Microtus agrestis*) at Lake Vyrnwy, Montgomeryshire, in 1936–9. *Philos. Trans. Roy. Soc. Lond., Ser. B* 236: 505–552.
Chitty, D. 1960. Population processes in the vole and their relevance to general theory. *Can. J. Zool.* 38: 99–113.
Chitty, D. 1967. The natural selection of self-regulatory behaviour in animal populations. *Proc. Ecol. Soc. Austral.* 2: 51–78.
Christian, J. J. 1978. Neurobehavioral endocrine regulation of small mammal populations. In D. P. Snyder (Ed.) *Populations of small mammals under natural conditions*. Special Publication Series, Pymatuning Laboratory of Ecology, Pittsburgh, Penn., pp. 143–158.
Collet, R. 1878. On *Myodes lemmus* in Norway. *Zool. J. Linn. Soc.* 13: 327–334.
Collet, R. 1895. *Myodes lemmus*, its habitats and migrations in Norway. *Forhandlingar Videskaplige-Selskapet Christiania* 3: 1–63.
Collet, R. 1911–12. *Norges pattedyr*. Kristiania.
Ehrström, Cr. 1852. Djurvandringar i Lappmarken och norra delen af Finland åren 1839 och 1850. *Notiser av Sällskapet Fauna och Flora* 2: 1–8.
Elton, C. 1924. Periodic fluctuations in the numbers of animals: their causes and effects. *Br. J. Exp. Biol.* 2: 119–163.

Elton, C., D. H. S. Davis, and G. M. Findlay. 1935. An epidemic among voles (*Microtus agrestis*) on the Scottish border in the spring of 1934. *J. Anim. Ecol.* 4: 277–288.
Erlinge, S. 1974. Distribution, territoriality and numbers of the weasel (*Mustela nivalis*) in relation to prey abundance. *Oikos* 25: 378–384.
Erlinge, S. 1975. Feeding habits of the weasel *Mustela nivalis* in relation to prey abundance. *Oikos* 26: 378–384.
Erlinge, S., G. Göransson, L. Hansson, G. H. Högstedt, O. Lidberg, I. H. Nilsson, T. Nilsson, T. von Schantz, and M. Sylvén. 1983. Predation as a regulating factor on small rodent populations in southern Sweden. *Oikos* 40: 36–52.
Fellman, J. 1848. *Bidrag till lappmarkens fauna*. Suomi, 1847.
Framstad, E., N. C. Stenseth, O. N. Bjørnstad, and W. Falck. 1997. Limit cycles in Norwegian lemmings: tensions between phase-dependence and density-dependence. *Philos. Trans. Roy. Soc. Lond., Ser. B* 264: 31–38.
Ginzburg, L. R. 1998. Assuming reproduction to be a function of consumption raises doubts about some popular predator–prey models. *J. Anim. Ecol.* 67: 325–327.
Gonzales, J.-P. and J.-M. Duplantier. 1999. The arenavirus and rodent co-evolution process: a global view of the theory. *In* J. F. Saluzzo and B. Dodet (Eds.) *Factors in the emergence and control of rodent-borne viral diseases (Hantaviral and Arenal diseases)*. Elsevier, Paris, pp. 39–42.
Grant, P. R. 1972. Interspecific competition among rodents. *Ann. Rev. Ecol. Syst.* 3: 79–106.
Hansen, T. F., N. C. Stenseth, and H. Henttonen. 1999a. Interspecific and intraspecific competition as causes of direct and delayed density dependence in a fluctuating vole population. *Proc. Natl. Acad. Sci. USA* 96: 986–991.
Hansen, T. F., N. C. Stenseth, H. Henttonen, and J. Tast. 1999b. Multiannual vole cycles and population regulation during long winters: an analysis of seasonal density dependence. *Am. Nat.* 154: 129–139.
Hanski, I. 1987. Populations of small mammals cycle—unless they don't. *Trends Ecol. Evol.* 2: 55–56.
Hanski, I. and H. Henttonen. 1996. Predation on competing vole species: a simple explanation of complex patterns. *J. Anim. Ecol.* 65: 220–232.
Hanski, I. and E. Korpimäki. 1995. Microtine rodent dynamics in northern Europe: parameterized models for the predator–prey interaction. *Ecology* 76: 840–850.
Hanski, I., L. Hansson, and H. Henttonen. 1991. Specialist predators, generalist predators and the microtine rodent cycle. *J. Anim. Ecol.* 60: 353–367.
Hanski, I., P. Turchin, E. Korpimäki, and H. Henttonen. 1993. Population oscillations of boreal rodents: regulation by mustelid predators leads to chaos. *Nature* 364: 232–235.
Hanski, I., H. Henttonen, and L. Hansson. 1994. Temporal variability of population density in microtine rodents: a reply to Xia and Boonstra. *Am. Nat.* 144: 329–342.
Hanski, I., H. Henttonen, E. Korpimäki, L. Oksanen, and P. Turchin. 2001. Small rodent dynamics and predation. *Ecology* 82: 1505–1520.
Hansson, L. 1983. Competition between rodents in successional stages of taiga forests: *Microtus agrestis* vs. *Clethrionomys glareolus*. *Oikos* 40: 258–266.
Hansson, L. 1985. Geographic differences in bank voles *Clethrionomys glareolus* in relation to ecogeographical rules and possible demographic and nutritive strategies. *Ann. Zool. Fenn.* 22: 319–328.
Hansson, L. 1987. An interpretation of rodent dynamics as due to trophic interactions. *Oikos* 50: 308–318.

Hansson, L. 1999. Intraspecific variation in dynamics: small rodents between food and predation in changing landscapes. *Oikos* 86: 159–169.

Hansson, L. and H. Henttonen. 1985a. Gradients in density variations of small rodents: the importance of latitude and snow cover. *Oecologia* 67: 394–402.

Hansson, L. and H. Henttonen. 1985b. Regional differences in cyclicity and reproduction in Clethrionomys spp: are they related? *Ann. Zool. Fenn.* 22: 277–288.

Hansson, L. and H. Henttonen. 1988. Rodent dynamics as community processes. *Trends Ecol. Evol.* 3: 195–200.

Hansson, L. and H. Henttonen. 1998. Rodent fluctuations in relation to seasonality in Fennoscandia and Hokkaido. *Res. Popul. Ecol.* 40: 127–129.

Haukisalmi, V. and H. Henttonen. 1990. The impact of climatic factors and host density on the long-term population dynamics of vole helminths. *Oecologia* 83: 309–315.

Haukisalmi, V. and H. Henttonen. 2000. Variability of helminth assemblages and populations in the bank vole *Clethrionomys glareolus*. *Ekol. Pol.* 48 (Suppl.): 219–231.

Heikkilä, J., A. Below, and I. Hanski. 1994. Synchronous dynamics of microtine rodent populations on islands in Lake Inari in northern Fennoscandia: evidence for regulation by mustelid predators. *Oikos* 70: 245–252.

Henttonen, H. 1985. Predation causing extended low densities in microtine cycles: further evidence from shrew dynamics. *Oikos* 45: 156–157.

Henttonen, H. 1986. *Causes and geographic patterns of microtine cycles*. Ph.D. thesis, University of Helsinki, Helsinki, Finland.

Henttonen, H. 1987. The impact of spacing behaviour in microtine rodents on the dynamics of least weasels *Mustela nivalis*—a hypothesis. *Oikos* 50: 366–370.

Henttonen, H. 2000. Long-term dynamics of the bank vole *Clethrionomys glareolus* at Pallasjarvi, northern Finnish taiga. *Ekol. Pol.* 48 (Suppl.): 87–96.

Henttonen, H. and I. Hanski. 2000. Population dynamics of small rodents in northern Fennoscandia. *In* J. N. Perry, R. H. Smith, I. P. Woiwod, and D. R. Morse (Eds.) *Chaos in real data*. Kluwer Academic, Dordrecht.

Henttonen, H. and L. Hansson. 1986. Synchrony and asynchrony between sympatric rodent species with special reference to *Clethrionomys*. *In* Henttonen (1986).

Henttonen, H. and A. Kaikusalo. 1993. Lemming movements. *In* N. C. Stenseth and R. A. Ims (Eds.) *The biology of lemmings*. Academic Press of the Linnean Society of London, London, pp. 157–186.

Henttonen, H. and H. Wallgren. 2000. Small rodent dynamics and communities in the birch forest zone of northern Fennoscandia. *In* F. E. Wielgolaski (Ed.) *Nordic mountain birch forest ecosystem*. UNESCO Man and Biosphere Series, Vol. 27, UNESCO, Paris and Parthenon Publishing Group, New York and London, pp. 261–278.

Henttonen, H., A. Kaikusalo, J. Tast, and J. Viitala. 1977. Interspecific competition between small rodents in subarctic and boreal ecosystems. *Oikos* 29: 581–590.

Henttonen, H., A. D. McGuire, and L. Hansson. 1985. Comparisons of amplitudes and frequencies (spectral analyses) of density variations in long-term data sets of *Clethrionomys* species. *Ann. Zool. Fenn.* 22: 221–227.

Henttonen, H., T. Oksanen, A. Jortikka, and V. Haukisalmi. 1987. How much do weasels shape microtine cycles in the northern Fennoscandian taiga? *Oikos* 50: 353–365.

Henttonen, H., V. Haukisalmi, A. Kaikusalo, E. Korpimäki, and K. Norrdahl. 1989. Long-term population dynamics of the common shrew *Sorex araneus* in Finland. *Ann. Zool. Fenn.* 26: 349–356.

Henttonen, H., L. Hansson, and T. Saitoh. 1992. Rodent dynamics and community structure: *Clethrionomys rufocanus* in northern Fennoscandia and Hokkaido. *Ann. Zool. Fenn.* 29: 1–6.

Hörnfeldt, B. 1994. Delayed density dependence as a determinant of vole cycles. *Ecology* 75: 791–806.

Ims, R. A. and H. Steen. 1990. Geographical synchrony in microtine population cycles: a theoretical evaluation of the role of nomadic avian predators. *Oikos* 57: 381–387.

Inchausti, P. and L. R. Ginzburg. 1998. Small mammals cycles in northern Europe: patterns and evidence for a maternal effect hypothesis. *J. Anim. Ecol.* 67: 180–194.

Jonasson, S., J. P. Bruant, F. Stuart Chapin III, and M. Andersson. 1986. Plant phenols and nutrients in relation to variations in climate and rodent grazing. *Am. Nat.* 128: 394–408.

Kaikusalo, A. and I. Hanski. 1985. Population dynamics of *Sorex araneus* and *S. caecutiens* in Finnish Lapland. *Acta Zool. Fenn.* 173: 283–285.

Kalela, O. 1962. On the fluctuations in the numbers of arctic and boreal small rodents as a problem of production biology. *Ann. Acad. Sci. Fenn., Ser. A IV.* 66: 1–38.

Kalela, O. and T. Koponen. 1971. Food consumption and movements of the Norwegian lemming in areas characterized by isolated fells. *Ann. Zool. Fenn.* 8: 80–84.

Keith, L. B., J. R. Cary, T. M. Yuill, and I. M. Keith. 1985. Prevalence of helminths in a cyclic snowshoe hare population. *J. Wildl. Dis.* 21: 233–253.

Klemola, T., K. Norrdahl, and E. Korpimäki. 2000. Do delayed effects of overgrazing explain population cycles in voles? *Oikos* 90: 509–516.

Korpimäki, E. 1993. Regulation of multiannual vole cycles by density-dependant avian and mammalian predation. *Oikos* 66: 359–363.

Korpimäki, E. and K. Norrdahl. 1989. Predation of Tengmalm's owls: numerical responses, functional responses and dampening impact on population fluctuations of microtines. *Oikos* 54: 154–164.

Korpimäki, E. and K. Norrdahl. 1991a. Do breeding nomadic avian predators dampen population fluctuations of small mammals? *Oikos* 62: 195–208.

Korpimäki, E. and K. Norrdahl. 1991b. Numerical and functional responses of kestrels, short-eared owls and long-eared owls to vole densities. *Ecology* 72: 814–826.

Korpimäki, E. and K. Norrdahl. 1998. Experimental reduction of predators reverses the crash phase of small-rodent cycles. *Ecology* 79: 2448–2455.

Korpimäki, E., M. Lagerström, and P. Saurola. 1987. Field evidence for nomadism in Tengmalm's owl *Aegolius funereus*. *Ornis Scand.* 18: 1–4.

Korpimäki, E., K. Norrdahl, and T. Rinta-Jaskari. 1991. Responses of stoats and least weasels to fluctuating vole abundances: is the low phase of the vole cycle due to mustelid predation? *Oecologia* 88: 552–561.

Krebs, C. J. 1996. Population cycles revisited. *J. Mammal.* 77: 8–24.

Krebs, C. J. and J. H. Myers. 1974. Population cycles in small mammals. *Adv. Ecol. Res.* 8: 268–400.

Lack, D. 1954. *The natural regulation of animal numbers*. Clarendon Press, Oxford.

Laine, K. and H. Henttonen. 1983. The role of plant production in microtine cycles in northern Fennoscandia. *Oikos* 40: 407–418.

Laine, K. M. and H. Henttonen. 1987. Phenolics/nitrogen ratios in the blueberry *Vaccinium myrtillus* in relation to temperature and microtine density in Finnish Lapland. *Oikos* 50: 389–395.

Lambin, X., S. J. Petty, and J. L. MacKinnon. 2000. Cyclic dynamics in field vole populations and generalist predation. *J. Anim. Ecol.* 69: 106–118.
Lawton, J. H. 1999. Are there general laws in ecology? *Oikos* 84: 177–192.
Leslie, P. 1948. Some further notes on the use of matrices in population dynamics. *Biometrika* 35: 213–245.
Lindström, E. R. and B. Hörnfeldt. 1994. Vole cycles, snow depth and fox predation. *Oikos* 70: 156–160.
Löfgren, O., B. Hörnfeldt, and B.-G. Carlsson. 1986. Site tenacity and nomadism in Tengmalm's owl (*Aegolius funereus*) in relation to cyclic food production. *Oecologia* 69: 321–326.
Magnus, O. 1555. *Historia de gentibvs....* Roma.
May, R. M. 1973. *Stability and complexity in model ecosystems.* Princeton University Press, Princeton, N.J.
May, R. M. 1991. The role of ecological theory in planning reintroduction of endangered species. *Symp. Zool. Soc. Lond.* 62: 145–163.
McGrady, S. J. and P. J. Morin. 1996. Disturbance and the species composition of rain pool microbial communities. *Oikos* 76: 93–102.
Moran, P. A. P. 1953. The statistical analysis of the Canadian lynx cycle. *Austr. J. Zool.* 1: 291–298.
Niederlehner, B. R. and J. Cairns Jr. 1994. Consistency and sensitivity of community level endpoints in microcosm tests. *J. Aquat. Ecosyst. Health* 3: 93–99.
Norrdahl, K. 1995. Population cycles in northern small mammals. *Biol. Rev.* 70: 621–637.
Norrdahl, K. and E. Korpimäki. 1995a. Effects of predator removal on vertebrate prey populations: birds of prey and small mammals. *Oecologia* 103: 241–248.
Norrdahl, K. and E. Korpimäki. 1995b. Mortality factors in a cyclic vole population. *Proc. Roy. Soc. Lond., Ser. B* 261: 49–53.
Oksanen, L. and T. Oksanen. 1992. Long-term microtine dynamics in north Fennoscandian tundra: the vole cycle and lemming chaos. *Ecography* 15: 226–236.
Oksanen, T., M. Schneider, U. Rammul, P. A. Hambäck, and M. Aunapuu. 1999. Population fluctuations of voles in North Fennoscandian tundra: contrasting dynamics in adjacent areas with different habitat composition. *Oikos* 86: 463–478.
Ostfeld, R. S., C. D. Canham, and S. R. Pugh. 1993. Intrisic density-dependent regulation of vole populations. *Nature* 366: 259–261.
Pimm, S. L. 1991. *Balance of nature?* University of Chicago Press, Chicago, Ill.
Pitelka, F. A. 1958. Some aspects of population structure in the short-term cycle of the brown lemming in northern Alaska. *Cold Spring Harbor Symp. Quant. Biol.* 22: 237–251.
Plyusnin, A., O. Vapalahti, Å. Lundkvist, H. Henttonen, and A. Vaheri. 1999. Hantaviruses in Europe: an overview. *In* J. F. Saluzzo and B. Dodet (Eds.) *Factors in the emergence and control of rodent-borne viral diseases (Hantaviral and Arenal diseases).* Elsevier, Paris, pp. 85–91.
Rosenzweig, M. L. and R. H. MacArthur. 1963. Graphical representation and stability conditions of predator–prey interaction. *Am. Nat.* 97: 209–223.
Royama, T. 1992. *Analytical population dynamics.* Chapman and Hall, London.
Saitoh, T. 1987. A time series and geographical analysis of population dynamics of the red-backed vole in Hokkaido, Japan. *Oecologia* 73: 382–388.
Sheftel, B. I. 1989. Long-term and seasonal dynamics of shrews in Central Siberia. *Ann. Zool. Fenn.* 26: 357–370.

Soveri, T., H. Henttonen, V. Haukisalmi, E. Rudbäck, A. Sukura, R. Tanskanen, R. Schildt, J. Husu, and J. Laakkonen. 2000. Disease patterns in field and bank vole populations during a cyclic decline in Central Finland. *Comp. Immunol. Microbiol. Infect. Dis.* 23: 73–89.

Steen, H. 1995. Untangling the causes of disappearance from a local population of root voles, *Microtus oeconomus*: a test of the regional synchrony hypothesis. *Oikos* 73: 65–72.

Stenseth, N. C. 1999. Population cycles in voles and lemmings: density dependence and phase dependence in a stochastic world. *Oikos* 87: 427–461.

Stenseth, N. C. and T. Saitoh. 1998. The population ecology of the vole *Clethrionomys rufocanus*. *Res. Popul. Ecol.* 40: 1–158.

Stenseth, N. C., O. N. Bjornstad, and T. Saitoh. 1996. A gradient from stable to cyclic populations of *Clethrionomys rufocanus* in Hokkaido, Japan. *Proc. Roy. Soc. Lond., Ser. B* 263: 1117–1126.

Taitt, M. J. and C. J. Krebs. 1985. Population dynamics and cycles. *In* R. H. Tamarin (Ed.) *Biology of new world microtus.* American Society Mammals, Special Publication no. 8, pp. 567–620.

Tanner, J. T. 1975. The stability and intrinsic growth rates of predator and prey populations. *Ecology* 56: 1835–1841.

Tast, J. and O. Kalela. 1971. Comparisons between rodent cycles and plant production in Finnish Lapland. *Ann. Acad. Sci. Fenn., Ser. A IV* 186: 1–14.

Turchin, P. 1993. Chaos and stability in rodent population dynamics: evidence from non-linear time-series analysis. *Oikos* 68: 162–172.

Turchin, P. and G. O. Batzli. 2001. Availability of food and the population dynamics of arviocoline rodents. *Ecology* 82: 1521–1534.

Turchin, P. and S. P. Ellner. 2000. Modelling time-series data. *In* J. N. Perry, R. H. Smith, I. P. Woiwod, and D. Morse (Eds.) *Chaos in real data.* Kluwer Academic, Dordrecht, pp. 33–48.

Turchin, P. and I. Hanski. 1997. An empirically based model for latitudinal gradient in vole population dynamics. *Am. Nat.* 149: 842–874.

Turchin, P. and I. Hanski. 2001. Contrasting alternative hypotheses about rodent cycles by translating them into parameterized models. *Ecol. Lett.* 4: 267–276.

Turchin, P. and R. S. Ostfeld. 1997. Effects of density and season on the population rate of change in the meadow vole. *Oikos* 78: 355–361.

Turchin, P., L. Oksanen, P. Ekerholm, T. Oksanen, and H. Henttonen. 2000. Are lemmings prey or predators? *Nature* 405: 493–598.

Virtanen, R., H. Henttonen, and K. M. Laine. 1997. Lemming grazing and the structure of snowbed plant community—a long-term experiment at Kilpisjärvi, in Finnish Lapland. *Oikos* 79: 155–166.

Whitney, P. and D. Feist. 1984. Abundance and survival of *Clethrionomys rutilus* in relation to snow cover in a forested habitat near College, Alaska. *In* J. F. Merritt (Ed.) *Winter ecology of small mammals.* Special Publication, Carnegie Museum of Natural History, Pittsburgh, Penn., pp. 113–120.

Ydenberg, R. C. 1987. Nomadic predators and geographical synchrony in microtine population cycles. *Oikos* 50: 270–272.

Yoccoz, N. G., N. C. Stenseth, H. Henttonen, and A.-C. Prévot-Julliard. 2001. Effects of food addition on the seasonal density-dependent structure of the bank vole *Clethrionomys glareolus* populations. *J. Anim. Ecol.* 70:713-720.

Ziegler, J. 1532. Quae intvs continentur. A new edition. *Svenska sällsk. Anthropology och geografi, Geografiska sectionens tidskrift* B1 (2): 1878.

4

Understanding the Snowshoe Hare Cycle through Large-scale Field Experiments

Stan Boutin, Charles J. Krebs, Rudy Boonstra, Anthony R. E. Sinclair, and Karen E. Hodges

4.1 Introduction

The 10-year cycles of the snowshoe hare and lynx seen in Hudson's Bay fur returns represent a classic example of cyclic population dynamics. Hare cycles have been the subject of time series analysis (Stenseth et al. 1998), population modeling (Royama 1992), and field experimentation (Keith and Windberg 1978, Krebs et al. 1986, Murray et al. 1997). However, only two studies have monitored hare populations in detail over at least one full cycle. The first of these was conducted in central Alberta, Canada, by Lloyd Keith and coworkers, and provided a detailed description of the demographic machinery driving changes in hare numbers (Keith et al. 1977, Cary and Keith 1979, Keith et al. 1984). From this came the "Keith hypothesis" that hare cycles are driven by a sequential two-stage trophic interaction with hare declines initiated by winter food shortages and exacerbated by predator numerical responses that lag hare numbers by 1–2 years (Keith 1983, 1990). Predators force hares to low numbers and recovery does not occur until predator densities reach their lowest levels.

The second long-term study of hare dynamics took place at Kluane Lake in the southwestern Yukon, Canada. The Kluane project began as an attempt to test the Keith hypothesis through single-factor manipulations of food supply and predation (Krebs et al. 1986, Sinclair et al. 1988, Smith et al. 1988). The first attempt failed to manipulate predators effectively, and plots containing food supplements were quickly overwhelmed by predators moving into the area. Consequently, the experiments failed to alter hare dynamics. Building on this experience, the second phase expanded the scale of experimental manipulations and developed an effective means of excluding predators from selected areas. The study also added an interaction treatment in which predators were excluded and food supplemented. These experiments were designed to test the roles of food supply, predation, and their potential interaction in the dynamics of snowshoe hares (Krebs et al. 1995). In this chapter we provide a synopsis of the key results obtained from these experiments and discuss how the results alter the current understanding of snowshoe hare dynamics.

4.2 Experimental Methods

The Kluane Boreal Forest Ecosystem Study took place in a 350 km^2 section of the Shakwak Trench stretching from the eastern shore of Kluane Lake, Yukon (see Boutin et al. 1995 for a more detailed description). The vegetation of the study area is typical of the Kluane sector of the boreal forest described by Rowe (1972). White spruce is the dominant tree in the region, and open and closed stands of spruce occupy the majority of the valley floor. The two lower zones are complex mosaics of forests of white spruce, stands of balsam poplar and aspen, and shrub-dominated areas of willow (mostly *Salix glauca*) and dwarf birch.

We applied the following treatments to 60–100 ha plots. We fertilized two 100 ha plots each spring with N–P–K distributed by airplane. On two 60 ha plots we broadcast laboratory rabbit chow ad lib throughout the year. Terrestrial predators were excluded from a single 100 ha plot by a fence composed of 5 cm wire mesh and electric wires. Holes were cut in the mesh to allow hares to pass through freely. We attempted to exclude avian predators from a 10 ha section of this plot using monofilament fishing line strung above the ground, but this proved unsuccessful. The final treatment was to combine the predator exclusion fence with food supplementation. This was done by placing a 60 ha food addition plot within a 100 ha fenced area. All experimental treatments were in place by fall 1988. Along with these treatments we monitored hare numbers on one to three control plots (60 ha each).

Estimates of snowshoe hare population density and increase rates were obtained by live-trapping in March and October. We trapped for three to seven nights per trapping grid per session but, to avoid hare weight loss, we never trapped for more than two nights in a row and spaced trapping episodes by two trap-free nights. Each trapping grid contained 86 Tomahawk traps (Tomahawk Live Trap Co., Tomahawk, Wisc.) placed along four rows with 30–60 m between traps, giving an effective grid size of 60 ha. We baited traps with alfalfa cubes and added apples or snow for moisture. Captured hares were ear-tagged, sexed, and weighed; in addition, the length of the right hindfoot and the reproductive condition of the animals were recorded (males: scrotal or abdominal testes, females: lactating, not lactating).

Survival estimates were obtained by collaring hares > 1000 g with 40 g radio transmitters containing mortality sensors. Radio signals were monitored every 1–2 days and all collars transmitting mortality signals were located to determine the cause of death using tracks, feces, bird castings, and methods of eating as clues to the identity of the predator. We were able to identify the species of predator in roughly half the deaths, and were able to separate avian from mammalian predation in another 10%.

Snowshoe hare reproduction and leveret survival were estimated using "maternity cages" (O'Donoghue 1994, Stefan 1998). Pregnant females were captured shortly before parturition and held in cages until they gave birth. Each leveret was ear-tagged, weighed, and sexed. Stillborn rates were calculated from leverets born dead, with necropsies confirming that these individuals died before birth (e.g., lungs not inflated, no signs of trampling). We radio-tagged about half of the leverets in each litter by gluing 2–2.5 g transmitters to their backs. Survival rates were determined by locating leverets daily until weaning ($\sim$ 4 weeks) or until the radios fell off (O'Donoghue 1994, Stefan 1998). We present natality results following Stefan (1998); these results differ from those in Krebs et al. (1995) and Boonstra et al. (1998b) because of reconsideration of the timing of the first litter.

4.3 Results

4.3.1 Patterns of Density Change

Peak hare densities on control plots occurred in 1989–90, with spring densities ranging from 1.6 to 2.0 hares/ha (figure 4.1). Low densities (0.01 to 0.10 hares/ha) were reached by the spring of 1993. Peak fall densities were higher than spring densities, with a maximum of 3.4 hares/ha on control site 3. Averaged across all control sites, the amplitude of the cycle was 18-fold for spring densities, and 30-fold if minimum spring densities are compared with maximum fall densities. We classified the phases of the cycle in the following way: peak = 1989 and 1990, decline = 1991 and 1992, low = 1993 and 1994, and increase = 1987, 1988, 1995, and 1996. Densities were highly variable among control plots during the increase phase but were much more consistent from fall 1990 to the low in 1993.

4.3.2 Treatment Effects on Average Densities

The experimental treatments all resulted in higher hare densities, with the largest increases occurring in the predator exclusion + food treatment. Averaged through the cycle, fertilization resulted in a 1.3-fold increase in density, food addition 3.1-fold, predator exclusion 2.0-fold, and predator exclusion + food addition 9.7-fold (figure 4.2). The fertilized sites had peak densities of 1.8 and 2.2 hares/ha in spring 1990 and low densities of 0.3 and 0.7 hares/ha in 1993 (figure 4.3a). Peak densities in the food additions sites

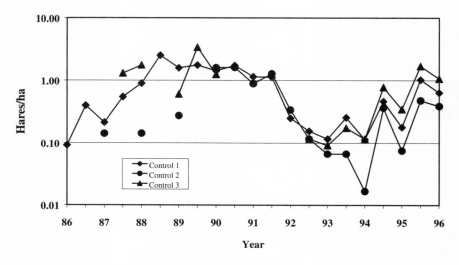

Figure 4.1 Spring and fall snowshoe hare densities on control sites through a population cycle. Densities were calculated using the average of Jolly–Seber and jackknife estimators for trapping sessions conducted in March and April (spring) and October (fall), and assuming an effective grid size of 60 ha. Control 3 was not trapped in 1986, 1987, 1991, or 1992; Control 2 was not trapped in 1986.

occurred in 1991, with densities (5.1 and 6.6 hares/ha) three to five times higher than on control sites (figures 4.2 and 4.3b). Low densities of 0.2 and 0.5 hares/ha occurred in 1993. In the predator exclosure, peak hare density (1.8 hares/ha) occurred in 1990 and a low of 0.2 hares/ha occurred in 1993 (figure 4.3c). Densities on this site then increased for 2 years and fell again in 1996. The largest treatment impact was observed in the predator enclosure + food treatment. Hares peaked at 6.1 hares/ha in 1990 and reached a low density of 1.0 hares/ha in 1993 (figure 4.3d). With the exception of fertilization, all treatments had their greatest impacts during the decline phase: relative to control sites, food addition sites had 3.8-fold higher densities, predator exclusion 2.4-fold, and the food + predator exclusion 14.4-fold during the decline (figure 4.2).

4.3.3 Treatment Effects on Population Change

The experimental manipulations also affected the timing and magnitude of changes in density. On control sites, hare populations declined steeply (yearly finite rates of growth = 0.197–0.556) from 1991 through to 1993 or 1994 (figure 4.1). There was a slight increase over the summer due to reproduction, but spring densities remained constant from 1993 to 1994. There followed an increase in 1994–95 and a greater increase in 1995–96. Hare populations on fertilizer treatments tracked control populations through the increase, peak, and decline (figure 4.3a), but then began to increase 1 year earlier.

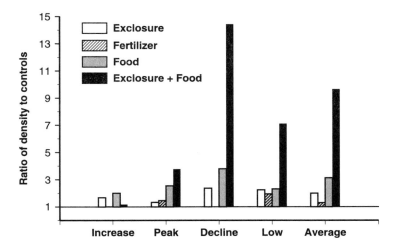

Figure 4.2 Ratio of hare population sizes on manipulated sites to mean control densities. Any value other than one indicates a treatment effect.

Populations on the food addition sites declined only from 1991 through 1993, but their decline rates and subsequent increase rates echoed those of control populations (figure 4.3b). Hares in the predator exclosure suffered a decline similar to the control hares, then increased for 2 years before an anomalous decline in 1995–96 (figure 4.3c). The population in the predator exclosure + food treatment remained high and constant from fall 1989 through fall 1992, before declining steeply for 1 year (yearly finite rate of increase = 0.299, figure 4.3d). Numbers remained relatively constant through to spring 1995, then increased rapidly.

4.3.4 Treatment Effects on Cycle Amplitude

The range in maximum amplitude of the cycle (maximum fall density to minimum spring density) on control grids was 22- to 94-fold. The predator exclusion treatment was just below this range (18-fold), while the fertilizer treatments were slightly above (105- and 167-fold). The amplitude on the first food treatment site was relatively high (39-fold), but that on the second was relatively low (11-fold). Predator exclusion + food addition reduced the amplitude of the cycle to 8-fold.

The effects of the various treatments on hare population density changes are summarized in table 4.1. The most striking effects occurred on the predator exclusion + food treatment, where densities were increased 10-fold on average, the major decline was delayed by a full year, and the magnitude of the decline was one third that on the control areas. Food addition increased the average density but had little effect on anything else. Fertilization of

74 Population Cycles

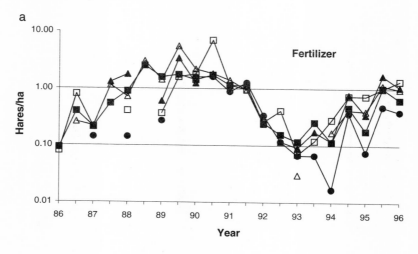

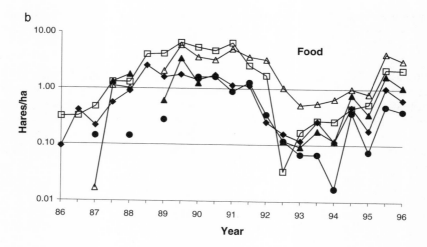

Figure 4.3 Snowshoe hare densities on treatment sites. Densities were calculated as in figure 4.1. Solid symbols represent control sites and open symbols represent experimental sites.

vegetation increased the amplitude of the cycle, and it was the only treatment that advanced the increase phase by 1 year. The predator exclosure produced low-phase densities that were higher than on control sites.

4.3.5 Treatment Effects on Reproduction

Snowshoe hares had their highest reproductive rates on control sites during the early increase phase (18.9 leverets/female per summer) and their lowest

Understanding the Snowshoe Hare Cycle 75

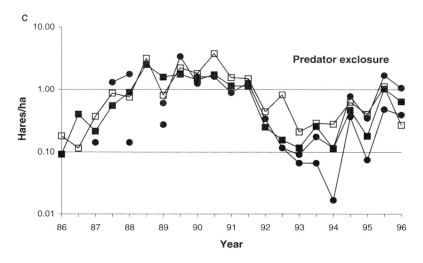

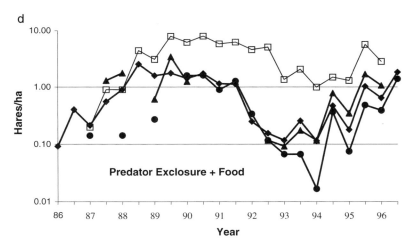

Figure 4.3 continued

reproduction during the decline (6.9 leverets/female per summer) (figure 4.4a). This difference was primarily due to the number of litters females produced, and secondarily to reduced sizes of the second and third litters (O'Donoghue and Krebs 1992, Stefan 1998). Snowshoe hares had four litters during the late low and early increase phases, three litters in the late increase and peak phases, and only two litters during the decline phase (O'Donoghue and Krebs 1992, Stefan 1998). Maximum sizes of the second and third litters occurred during the increase and minimum sizes during the decline, whereas the first litter remained constant throughout the cycle.

Table 4.1 Summary of changes in hare numbers on experimental treatments

Treatment	Average Density Difference[1]	Peak Density in Spring (hares/ha)	Lowest Density (hares/ha)	Cycle Amplitude[2]	Initiation of Major Decline[3]	No. of Years of Major Decline	Year of Increase[4]
Controls	—	1.6–2	0.01–0.1	22–94	1991–92	2–4	1994
Fertilizer	1.3	1.8–2.2	0.3–0.7	105–167	1991–92	3	1993
Food	3.1	5.1–6.6	0.2–0.5	11–39	1991–92	2	1994–95
Predator exclusion	2.0	1.8	0.2	18	1991–92	2	1994
Predator exclusion + food	9.7	6.1	1.0	8	1992–93	1	1995

1. Calculated as the ratio of treatment density/control density.
2. Calculated as the ratio of maximum peak density/minimum low density.
3. Major decline defined as a yearly finite rate of increase < 0.70.
4. Year when yearly finite growth rates > 2.

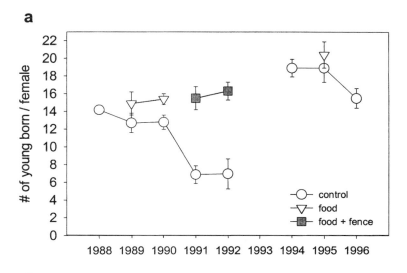
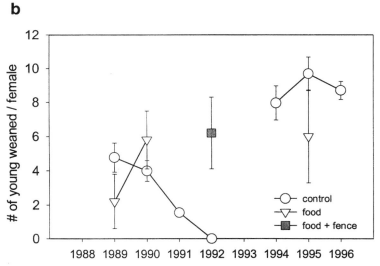

Figure 4.4 Snowshoe hare reproductive output and recruitment through the cycle. (a) Values are total young per female per summer, calculated from pregnancy rates and mean litter size per litter group. Litter 4 pregnancy rates and litter sizes for 1994 and 1995 were estimated based on values for last litters in other years, because the fourth litter was not measured directly. Pregnancy rates for control hares in 1988 and predator exclosure + food hares in 1991 and 1992 were based on averages for hares in 1989 and 1990. Standard errors were calculated from litter sizes. (b) Recruitment to 30 days. Values are calculated from reproductive output and survival to 30 days. For the food addition grids in 1989, there was no survival estimate for litter 1, so we used the conservative value of survival of the second litter (0.15). This value is conservative because in most years hares in second litters had lower survival than hares in first litters.

Our ability to compare reproductive rates between control and experimental areas depended on graduate student projects that focused primarily on the effects of food on reproduction (O'Donoghue and Krebs 1992, Stefan 1998). During the peak and early increase phases, hares on food addition and control sites had similar reproductive output. We do not have reproductive data from food addition areas during the population decline. However, we detected a difference between predator exclosure + food and control areas during the decline: hares in the treatment had three litters in 1991 and 1992, while control females had only two litters. Similarly, hares in the predator exclosure + food site had large second litters in both years (mean 7.6 leverets per litter), while hares on control areas had smaller second litters (mean 4.2) than during the increase and peak. The net effect is that control females had 12.8 leverets in 1990 and 7.0 in 1992, whereas females in the predator exclosure + food site had 16.3 leverets in 1992.

We also detected differences in stillborn rates between food addition and control sites. On control areas, the highest stillborn rate (30%) occurred in litter 2 in 1991. In each year, the first litter had the lowest stillborn rate (<8%). Food addition resulted in stillborn rates approximately double those of hares from control areas, with the highest rates for litter 3 in 1989 (45%) and 1990 (24%).

Leveret survival to 30 days varied through the cycle, both among litter groups and with experimental treatment, but second litter leverets typically had the lowest survival rates (table 4.2). Leveret survival was highest during the increase phase, lowest during the decline, and quite variable during the peak phase. Leverets on control and food addition sites had similar survival rates, with the exception of 1995 (early increase), when hares on food addition sites actually had lower survival rates (by 12–44%). Again, we detected large differences between control and predator exclosure + food addition sites; for example, no radio-tagged leverets survived in 1992 (although we did capture juvenile hares in the later trapping sessions), while 21–43% survived in the predator exclosure + food addition site. Notice that survival rates in the predator exclosure + food addition site were still low in comparison with those on control areas during the increase phase (37–71% survival). Across the entire cycle, 70% of the leverets killed by predators died in their first week of life and 23% died in the second week.

The combination of changes in reproductive output and leveret survival resulted in the most leverets being weaned per female during the early increase phase (9.7 in 1995), while the number declined throughout the population decline (figure 4.4b). Food addition did not have a consistent effect on weaning success: relative to control sites, females on food addition sites weaned more young during the peak, but fewer during the increase phase. Females we observed on control plots in 1992 failed to wean any young, while those on the predator exclosure + food addition plots weaned 6.2 young in that year.

Table 4.2 Preweaning survival of snowshoe hares. Kaplan–Meier techniques were used to estimate survival (±SE) of radio-tagged leverets from birth until 30 days old; hares were typically weaned between 4 and 5 weeks of age. The numbers of hares monitored are given in parenthesis. Data are from O'Donoghue (1994) and Stefan (1998)

	1989	1990	1991	1992	1994	1995	1996
Control							
Litter 1	0.73 ± 0.13 (12)	0.27 ± 0.13 (11)	0.47 ± 0.12 (17)	0 (9)	0.50 ± 0.22 (27)	0.71 ± 0.11 (19)	0.61 ± 0.08 (39)
Litter 2	0.22 ± 0.09 (23)	0.13 ± 0.05 (41)	0 (21)	0 (4)	0.37 ± 0.08 (46)	0.61 ± 0.07 (49)	0.54 ± 0.06 (66)
Litter 3	0.18 ± 0.12 (11)	0.51 ± 0.11 (24)	—	—	0.66 ± 0.07 (50)	0.60 ± 0.09 (31)	0.56 ± 0.07 (61)
Food							
Litter 1	—	0.45 ± 0.10 (31)				0.49 ± 0.15 (14)	
Litter 2	0.15 ± 0.07 (28)	0.15 ± 0.07 (33)				0.17 ± 0.08 (22)	
Litter 3	0.15 ± 0.13 (15)	0.57 ± 0.11 (27)				0.48 ± 0.10 (28)	
Predator exclosure + food							
Litter 1				0.21 ± 0.11 (26)			
Litter 2				0.37 ± 0.09 (39)			
Litter 3				0.43 ± 0.09 (36)			

4.3.6 Treatment Effects on Adult Survival Rates

Adult survival of snowshoe hares was highest during the increase phase (30 day survival of 0.91 in 1988–89 and 1995–96, giving an annual survival rate of 32%) and lowest during the decline (30 day survival of 0.64 during 1991–92, giving an annual rate of 0.5%) (figure 4.5). Hares on control sites had 30 day survival rates greater than 0.90 only in the increase phase (1988–89 and 1995–96). In clear contrast, monthly survival was lower than 0.90 in only 1 year within the predator exclosures. The lowest survival in the predator exclosure site was 0.83 in 1991–92, and the lowest survival in the predator exclosure + food addition site was 0.89 in 1992–93. These years marked the fastest rates of decline in these areas. Fertilization did not affect snowshoe hare survival rates and food addition did not have a consistent effect: hares on food addition sites had quite low survival during the decline (0.70 and 0.62 during 1991–92 and 1992–93, respectively), but slightly higher survival than control hares during other cyclic phases.

The predator exclusion fences had holes cut in them to allow hares to go through the fences. During the entire cycle, 31% of the radio-collared hares from the predator exclosure site and 35% of those from the predator exclosure + food addition site died outside the fences. Although we excluded these hares from our calculations, we think this movement through the fences contributed to the population declines on these sites because the percentage of radio-collared hares leaving the fenced areas increased as the decline proceeded (figure 4.6). Emigration rates were particularly high from the predator exclosure in 1992–93, and the highest rates of emigration from the predator exclosure + food grid occurred in 1993–94.

4.3.7 Causes of Death

We observed 1156 snowshoe hare mortalities during the population cycle. In all years and on all treatments, most hares died from predation (78%). If we exclude cases in which the cause of death could not be determined, then this figure rises to 95%. This general pattern also holds within fenced sites, with 79% of all deaths and 91% of deaths due to identifiable causes caused by predators. Some hares died from starvation or injury during the peak and decline years on control sites, but deaths from these causes were not observed during the low phase. Some hares on the predator exclosure + food addition site died from causes other than predation in all but 1 year during the low phase.

Of all hare deaths we recorded, 15% could be identified only as predator-caused and another 15% were unidentifiable. These deaths are expected to be distributed in a similar fashion to those we could identify, although great horned owl kills may be harder to identify because of a lack of diagnostic characters. Of all the adult deaths, only 18 were due to wolves, marten, wolverine, weasels, bald eagles, Harlan's hawks, and hawk owls. All the others were caused by lynx, coyotes, goshawks, and great horned owls. In the

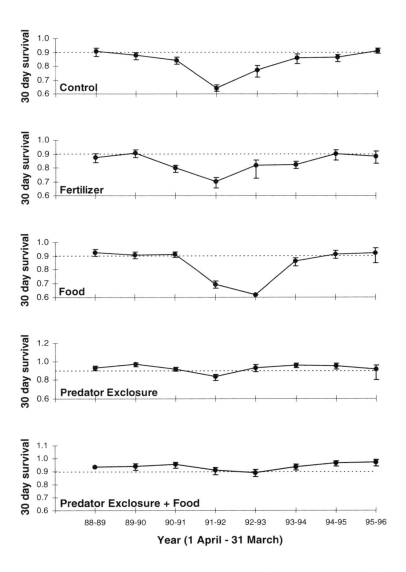

Figure 4.5 Adult snowshoe hare survival. Values are 30-day survival and 95% confidence interval (based on equation 3, Pollock et al. 1989); the line at 0.90 is given for comparison among treatments. A 30-day survival of 0.90 corresponds to an annual survival rate of 27.8%. Survival estimates were calculated from 1 April through 31 March. The point for fertilizer grids in 1995–96 is based on the period 1 April to 29 February, since many radio-collars were removed in the spring trapping census, making March survival estimates suspect. The 1995–96 estimate for predator exclosure + food is based on the period 1 April to 8 November, since a coyote inside the fence inflicted heavy mortality thereafter. Within each year, sample sizes were 42–168 hares (control), 26–131 (fertilizer), 44–114 (food), 25–113 (predator exclosure), and 35–94 (predator exclosure + food).

82 Population Cycles

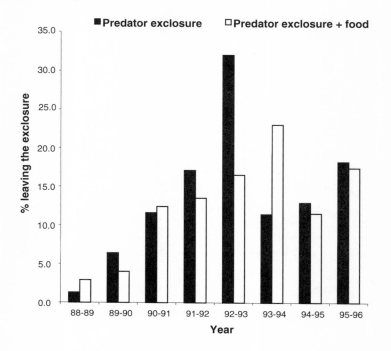

Figure 4.6 Percentage of radio-collared hares leaving the fences within each year, 1 April to 31 March. Through the entire cycle 31.2% or 62/199 hares left the predator exclosure and 35.4% or 84/237 hares left the predator exclosure + food area. In the figure, the percentages are lower because the values are calculated as (no. dead outside)/(no. radio-collared per year), and many radio-collared hares lived more than 1 year.

unfenced treatments (controls, fertilizer, food addition), approximately 65–75% of hare deaths were caused by mammalian predators, primarily lynx and coyotes. In contrast, inside the predator exclusion fences (which were designed to keep out mammalian predators), 70–80% of hare deaths were caused by raptors, especially great horned owls and goshawks. The fences did not exclude all mammalian predators, and across all years lynx killed 12.4% and 7.8% of radio-collared hares in the predator exclosure and predator exclosure + food sites, respectively. During winter 1995–96, a coyote got into the predator exclosure + food addition fence for about 3 months and was responsible for half of the hare deaths on this treatment during the entire year.

Predation mortality showed a seasonal pattern on control sites; for example, 65% of hares killed by coyotes died in October and November (coyotes cached 55% of these hares whole), while 70–75% of lynx, goshawk, and great horned owl kills occurred between December and May (figure 4.7).

Leverets were mainly killed by smaller predators (O'Donoghue 1994, Stefan 1998) and no radio-tagged leverets were killed by lynx or coyotes.

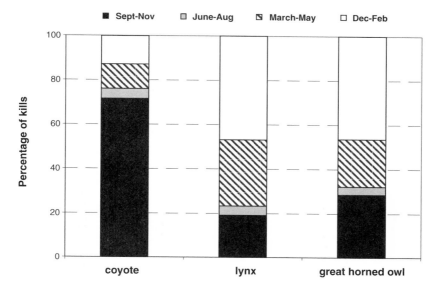

Figure 4.7 Seasonality of hare mortalities. Each bar indicates all the deaths of control hares for which cause of death could be determined for the period 1 April 1988 to 31 March 1996. Sample sizes for coyotes, lynx, and great horned owl were 109, 47, and 28, respectively.

Figure 4.8 shows the sources of mortality of radio-tagged leverets on control areas for whom fates were known. Red squirrels killed 40% of the sample during the peak, fewer during the decline, and an increasing number during the hare population increase. In 1991, 70% of the radio-tagged leverets appeared to be abandoned and to die of exposure. Abandonment also occurred in 1992 but was virtually absent in other years. Great horned owls, goshawks, boreal owls, Harlan's hawks, and weasels also killed leverets with the highest combined mortality from these predators (18.5%) occurring in 1994.

4.4 Discussion and Conclusions

4.4.1 The Demographic Mechanics of the 1986–96 Kluane Hare Cycle

Our study of the hare cycle on control grids indicates that the population increase was stopped by a combination of reduced reproduction and deteriorating juvenile and adult survival. Reproductive changes prior to the population decline were small and primarily due to the absence of a fourth litter. We were surprised to find, however, that red squirrels killed as many as

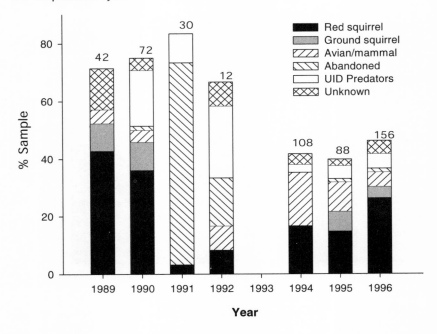

Figure 4.8 Causes of mortality of radio-tagged leverets on control areas. The values are percentages of the total number of radio-tagged individuals for which fate could be determined; remaining leverets survived. Leverets were monitored from birth to 30 days of age. Sample sizes are shown above each bar. The "avian/mammal" category includes great horned owl, goshawk, boreal owl, Harlan's hawk, and short-tailed weasel. The "UID predators" category is for unidentified predation deaths where there were insufficient signs to identify the species of predator.

50% of the leverets produced during the peak. This loss, combined with a small decrease in adult survival, led to a slow decline that became more severe with further reductions in recruitment and catastrophic adult mortality. The proximate cause of mortality of postweaning hares was predation, with lynx and coyotes responsible for roughly two thirds of all deaths. Adult survival recovered in the low phase but the population did not begin to increase until recruitment had recovered.

Our study represents only the third time that researchers have described hare demography over a complete cycle. For the most part, the results among studies are consistent, but our study establishes predators as the predominant force through all but the increase phase of the cycle. Predators were also the primary cause of adult hare mortality at Rochester, Alberta (Keith et al. 1984) and in the previous Kluane cycle (Boutin et al. 1986). The numerical (Rohner 1996, O'Donoghue et al. 1997) and functional responses (O'Donoghue et al. 1998) of lynx, coyotes, and owls were greater than in previous studies, and the increase in kill rates over the cycle was roughly 15 to 20-fold. Our estimates of the total impact on hares by coyote and lynx (for which both numerical and functional responses could be calculated) ranged from 9–13% during the increase phase, to 21% at the peak, 21–40% during the decline, and 15–47% during the low phase (O'Donoghue et al. 1997, 1998). These estimates are approximately double those from Alberta (Keith et al. 1977).

Although the rise and decline in adult hare mortality from predators lagged behind hare densities by 1–2 years (O'Donoghue et al. 1997, 1998), giving the delayed negative feedback needed to drive endogenous cycles, it is still not clear if predators are capable of stopping the increase phase. We did observe a reduction in reproductive output (no fourth litter produced) at the peak, but this reduction was generally less than that recorded at Rochester (Cary and Keith 1979). Community models of the boreal food web suggest that even large reductions in reproductive output during the peak could not initiate the decline phase, and that high mortality rates are necessary to achieve this transition (Ruesink and Hodges 2001, Ruesink et al. in review). Our results suggest that starvation is not involved in initiating the decline, and that almost all adult mortality during this transition is due to predators. In other words, our results do not support the starvation component of the Keith hypothesis (Keith 1983, 1990). Furthermore, our results indicate that the killing of leverets by red and ground squirrels could reduce recruitment enough to enable predators on adult hares to halt the increase phase. For this scenario to be true, squirrels would need to have a type III functional response to leverets. Our data (figure 4.8) provide some evidence that this may be the case, at least for red squirrels (see also Stefan 1998).

4.4.2 What the Experimental Treatments Revealed

As with all experimental studies, our work suffers from shortcomings that must be considered when interpreting the results. We decided to trade off replication for large-scale treatments, with the result that we did not replicate the predator exclusion treatments. We did have multiple controls to provide some estimate of natural variability, but the experimental effects still had to be large to be convincing. Further, we could not exclude avian predators, which were responsible for roughly one third of all hare deaths on control sites. In addition, we allowed hares to move freely through the predator fences because we did not want complications caused by "fence effects." Keeping these shortcomings in mind, we provide the following interpretation of the experimental results below.

4.4.2.1 Changes in Population Density

Fertilization, food addition, and predator reduction all resulted in increased hare densities relative to control sites. The effect of fertilization was, however, much more striking on vegetation than on hares (Turkington et al. 1998). The joint manipulation of predator exclosure + food addition had more than additive effects on snowshoe hare densities. For example, simple predator exclusion increased hare densities 2.4-fold during the decline, food addition increased densities 3.8-fold, while the joint manipulation increased hare densities 14.4-fold. This result indicates that food and predation interact in their effects on hare population dynamics.

This interpretation is supported by the observation that hare populations on all sites (except predator exclosure + food addition) experienced 2 years of severe population decline and cyclic amplitudes similar to those on control sites. In contrast, the manipulation of food and predators resulted in substantially higher hare densities, a 1-year delay and reduction in the decline phase, and a reduced cycle amplitude. Even in this case, however, the snowshoe hare cycle was still evident.

4.4.2.2 Demographic Effects

Keith (1983, 1990) concluded that winter food supplies influenced snowshoe hare reproduction and perhaps survival. Our results do not support Keith's proposition because hare survival and population change did not differ between control sites and fertilization sites, on which plant biomass and protein content were much higher. The addition of rabbit food also contributed biomass and high protein (approximately three to four times as much protein as in the twigs that hares normally eat), but did not affect the reproductive output of hares (O'Donoghue and Krebs 1992, Stefan 1998). Although food addition resulted in slightly earlier reproduction and somewhat larger litter sizes in some years, it did not affect the number of litters born per year and hares on food sites had higher stillborn rates. In addition, food supplementation did not affect the survival of leverets or juvenile hares. In peak years, hares on food addition sites had slightly higher survival than those on control areas, but this pattern was reversed during the decline when predators appeared to concentrate their activities on food addition sites, presumably because of the higher hare densities (apparently caused by the movement of hares onto these sites; see Boutin 1984, Krebs et al. 1986, Hodges et al. 2001).

Predator exclusion resulted in higher adult survival rates but, despite this effect, the cyclic patterns in survival and density persisted. Although we lack information on hare reproduction for the predator exclosure, it appears to have been similar to that on control areas. We arrived at this conclusion because both control and predator exclosure sites showed similar numerical changes through the cycle, while the predator exclosure + food addition site showed various irregularities. Adult survival rates on the two predator exclosures were similar, suggesting that the change in adult survival was not enough to cause the numerical anomalies shown by the predator exclosure + food addition site. However, this treatment had higher reproductive output and leveret survival than the controls during the decline phase, suggesting that these demographic elements led to the differences. The lack of similar irregularities on the predator exclosure indicates that recruitment was similar to control areas.

The single-factor manipulations suggest that changes in hare recruitment may depend on both food availability and predator pressure. Hares in the two-factor manipulation were able to maintain three litters and large litter sizes in years when hares on other sites were experiencing rapid population declines. This high reproduction and survival should have enabled the population to remain at high densities during the cyclic decline elsewhere, but a large proportion of hares left when densities outside the fence declined. This emigration contributed to the population decline seen in this treatment (figure 4.6), as did the relatively low survival rates of leverets.

We think that hare emigration through fences was triggered by the availability of natural food. Despite ad lib food supplementation on predator exclusion + food treatment, hares continued to utilize natural vegetation to the point that many of the shrubs and some of the small spruce trees were ring-barked (Hodges 1998). Hares may have ranged outside the fence in search of live natural vegetation. The same may be true in the predator exclosure. Although browsing levels there appeared similar to control areas, hare densities were higher, and movements may have increased during the peak and decline, thereby increasing hare exposure to predators. Predators also became adept at using the fence to capture hares. We observed a number of instances where hares being pursued by lynx outside the fence could not find a hole to enter the fence, despite the fact that we had created thousands of entry points.

4.4.3 Causes of the Hare Cycle

Previous models describing the snowshoe hare cycle have emphasized single-factor explanations, such as food–hare interactions (Bryant 1981, Fox and Bryant 1984) or hare–predator interactions (Trostel et al. 1987). Even Keith's model (1983, 1990) suggests that the transition from peak to decline phase is primarily due to a food–hare interaction, while the transition from low to increase phase is due to a hare–predator interaction. Our results strongly indicate that these single-factor or sequential models are inadequate. Predators clearly have a major impact on hares, especially during the cyclic decline, but reproductive changes also appear necessary for transitions between phases to occur, as shown by the joint manipulation of food and predators.

From this experimental work, we obtain the following demographic understanding of the snowshoe hare cycle. Obviously, the cyclic peak turns into a decline when recruitment no longer exceeds mortality. Our results suggest that this transition is due to the combined effect of reduced reproduction and increased mortality from predators. Furthermore, we did not find any evidence that malnutrition contributes to the population decline. Reducing the impact of mammalian predators resulted in increased adult survival, but does not appear to change reproductive output, and does not alter the cyclic pattern appreciably. Increasing food does not seem to affect either recruitment or mortality enough to change the cycle. The shift from the low phase to the increase phase requires reproduction to exceed mortality and, at least during this population cycle, this shift appeared to be primarily due to an increase in reproduction. This result is in contrast to the idea that delayed numerical responses of predators to hare density are necessary to "release" hares from the cyclic low. In theory, either reduced predation or increased reproduction would allow hares to enter the increase phase, and the actual balance may vary among cycles.

The next step in understanding the hare cycle is to determine why the reproductive and mortality rates change, especially since predator abundance and habitat structure may lead to complex interactions between food and predator effects on hares. Several promising approaches focus on the behavioral and physiological responses of hares to food and predators (Wolff 1980, 1981, Sievert and Keith 1985, Sovell 1993, Hik 1994, Murray et al. 1997, 1998, Boonstra et al. 1998b, Hodges 1998). In particular, changes in hare habitat-use patterns, movements, or foraging behavior in response to predator pressure could affect their survival and fecundity (Dolbeer and Clark 1975, Wolff 1981, Hik 1994, Hodges 1998). Furthermore, changes in stress physiology and parasite loads have been observed through portions of the cycle (Sovell 1993, Sovell and Holmes 1996, Murray et al. 1997, 1998, Boonstra et al. 1998a,b) and these physiological changes may have demographic repercussions.

Although the predator exclusion + food treatment altered the hare cycle substantially, the cycle was still apparent. Is there any treatment that can stop the cycle completely? To date, all experimental manipulations of food and predators have been initiated during the increase phase and the result has been a substantial increase in density by the time the peak is reached. It would be interesting to begin supplying food at the hare peak to try to prevent this buildup in density and to see if the decline could be delayed or eliminated. We also suspect that determining the factors causing the reproductive shifts would greatly aid our understanding of cyclic dynamics. So far, the most promising approaches in that direction link stress physiology, food availability, and predator avoidance behavior.

ACKNOWLEDGMENTS AND COMMENTS

We wish to thank the many people who contributed to collection of the data presented in this chapter. Thanks also to Ainsley Sykes for her editorial assistance and to Alan Berryman for the opportunity to contribute to this book. In his review of our chapter, Alan was somewhat dissatisfied with our strict adherence to "classic experimental ecology." He suggests that time series analysis and modeling could be used to indicate the types of experiments needed to distinguish between competing hypotheses and to provide an organizational structure for field ecology. We disagree only in terms of emphasis. The Kluane experiments were based on our best attempt to test a number of competing hypotheses and our results have clearly refuted some of these while providing support for others (see also Krebs et al. 2001). Although we do not discuss time series analysis in our chapter, we direct readers to Stenseth et al. (1998) for some ideas on how the Kluane experiments link to this sort of approach. The most important result was that the single-factor experiments of food addition or predator exclusion had little effect on the cycle, whereas the food addition–predator exclusion experiment came very close to stopping the cycle (delaying the decline and reducing the amplitude 4–5-fold). This supports some form of three-trophic-level hypothesis, as suggested by time series and loop analyses (Berryman 1981, Stenseth et al. 1998, Dambacher et al. 1999). Identification of the specific architecture of this three-trophic-level system will probably have to await the construction of mechanistic models that explain both the natural population cycle and the experimental results.

REFERENCES

Berryman, A. A. 1981. *Population systems: a general introduction*. Plenum Press, New York.

Boonstra, R., C. J. Krebs, and N. C. Stenseth. 1998a. Population cycles in small mammals: the problem of explaining the low phase. *Ecology* 79: 1479–1488.

Boonstra, R., D. Hik, G. R. Singleton, and A. Tinnikov. 1998b. The impact of predator-induced stress on the snowshoe hare cycle. *Ecol. Monogr.* 79: 371–394.

Boutin, S. 1984. Effect of late winter food addition on numbers and movements of snowshoe hares. *Oecologia* 62: 393–400.

Boutin, S., C. J. Krebs, A. R. E. Sinclair, and J. N. M. Smith. 1986. Proximate causes of losses in a snowshoe hare population. *Can. J. Zool.* 64: 606–610.

Boutin, S., C. J. Krebs, R. Boonstra, M. R. T. Dale, S. J. Hannon, K. Martin, A. R. E. Sinclair, J. N. M. Smith, R. Turkington, M. Blower, A. Byrom, F. I. Doyle, C. Doyle, D. Hik, L. Hofer, A. Hubbs, T. Karels, D. L. Murray, M. O'Donoghue, C. Rohner, and S. Schweiger. 1995. Population changes of the vertebrate community during a snowshoe hare cycle in Canada's boreal forest. *Oikos* 74: 69–80.

Bryant, J. P. 1981. The regulation of snowshoe hare feeding behaviour during winter by plant antiherbivore chemistry. *In* K. Myers and C. D. MacInnes (Eds.). *Proceedings of the World Lagomorph Conference*, University of Guelph, Guelph, Ont., pp. 720–731.

Cary, J. R. and L. B. Keith. 1979. Reproductive change in the 10-year cycle of snowshoe hares. *Can. J. Zool.* 57: 375–390.

Dambacher, J. M., H. W. Li, J. O. Wolff, and P. A. Rossignol. 1999. Parsimonious interpretation of the impact of vegetation, food, and predation on snowshoe hare. *Oikos* 84: 530–532.

Dolbeer, R. A. and W. R. Clark. 1975. Population ecology of snowshoe hares in the central Rocky Mountains. *J. Wildl. Manage.* 39: 535–549.

Fox, J. F. and J. P. Bryant. 1984. Instability of the snowshoe hare and woody plant interaction. *Oecologia* 63: 128–135.

Hik, D. S. 1994. *Predation risk and the 10-year snowshoe hare cycle*. Ph.D. thesis, University of British Columbia, Vancouver, B.C.

Hodges, K. E. 1998. *Snowshoe hare demography and behaviour during a cyclic population low phase*. Ph.D. thesis. University of British Columbia, Vancouver, B.C.

Hodges, K. E., C. J. Krebs, D. S. Hik, C. I. Stefan, E. A. Gillis, and C. E. Doyle. 2001. Snowshoe hare demography. *In* C. J. Krebs, S Boutin, and R. Boonstra (Eds.)

Ecosystem dynamics of the boreal forest. Oxford University Press, New York, pp. 141–178.
Keith, L. B. 1983. Role of food in hare population cycles. *Oikos* 40: 385–395.
Keith, L. B. 1990. Dynamics of snowshoe hare populations. *In* H. H. Genoways (Ed.) *Current mammalogy.* Plenum Press, New York, pp. 119–195.
Keith, L. B. and L. A. Windberg. 1978. A demographic analysis of the snowshoe hare cycle. *Wildl. Monogr.* 58: 1–70.
Keith, L. B., A. W. Todd, C. J. Brand, R. S. Adamcik, and D. H. Rusch. 1977. An analysis of predation during a cyclic fluctuation of snowshoe hares. *Proc. Int. Congr. Game Biol.* 13: 151–175.
Keith, L. B., J. R. Cary, O. J. Rongstad, and M. C. Brittingham. 1984. Demography and ecology of a declining snowshoe hare population. *Wildl. Monogr.* 90: 1–43.
Krebs, C. J., B. S. Gilbert, S. Boutin, A. R. E. Sinclair, and J. N. M. Smith. 1986. Population biology of snowshoe hares. I. Demography of food-supplemented populations in the southern Yukon, 1976–84. *J. Anim. Ecol.* 55: 963–982.
Krebs, C. J., S. Boutin, R. Boonstra, A. R. E. Sinclair, J. N. M. Smith, M. R. T. Dale, K. Martin, and R. Turkington. 1995. Impact of food and predation on the snowshoe hare cycle. *Science* 269: 1112–1115.
Krebs, C. J., R. Boonstra, S. Boutin, and A. R. E. Sinclair. 2001. What drives the 10-year cycle of snowshoe hares? *Bioscience* 51 (1): 25–35.
Murray, D. L., J. R. Cary, and L. B. Keith. 1997. Interactive effects of sublethal nematodes and nutritional status on snowshoe hare vulnerability to predation. *J. Anim. Ecol.* 66: 250–264.
Murray, D. L., L. B. Keith, and J. R. Cary. 1998. Do parasitism and nutritional status interact to affect production in snowshoe hares? *Ecology* 79: 1209–1222.
O'Donoghue, M. 1994. Early survival of juvenile snowshoe hares. *Ecology* 75: 1582–1592.
O'Donoghue, M. and C. J. Krebs. 1992. Effects of supplemental food on snowshoe hare reproduction and juvenile growth at a cyclic population peak. *J. Anim. Ecol.* 61: 631–641.
O'Donoghue, M., S. Boutin, C. J. Krebs, and E. J. Hofer. 1997. Numerical responses of coyotes and lynx to the snowshoe hare cycle. *Oikos* 80: 150–162.
O'Donoghue, M., S. Boutin, C. J. Krebs, G. Zuleta, D. L. Murray, and E. J. Hofer. 1998. Functional responses of coyotes and lynx to the snowshoe hare cycle. *Ecology* 79: 1193–1208.
Pollock, K. H., S. R. Wintersten, C. M. Bunck, and P. D. Curtis. 1989. Survival analysis in telemetry studies: the staggered entry design. *J. Wildl. Manage.* 53: 7–15.
Rohner, C. 1996. The numerical response of great horned owls to the snowshoe hare cycle: consequences of non-territorial 'floaters' on demography. *J. Anim. Ecol.* 65: 359–370.
Rowe, J. S. 1972. Forest regions of Canada. *Can. Forest Serv. Publ.* 1300: 1–172.
Royama, T. 1992. *Analytical population dynamics.* Chapman and Hall, London.
Ruesink, J. L. and K. E. Hodges. 2001. Trophic mass balance models of the Kluane boreal forest ecosystem. *In* C. J. Krebs, S. Boutin, and R. Boonstra (Eds.), *Ecosystem dynamics of the boreal forest: the Kluane project.* Oxford University Press, New York, pp. 436–490.
Ruesink, J. R., K. E. Hodges, and C. J. Krebs. Mass-balance analyses of boreal forest population cycles: merging demographic and ecosystem approaches. *Ecosystems* (accepted).

Sievert, P. R. and L. B. Keith. 1985. Survival of snowshoe hares at a geographic range boundary. *J. Wildl. Manage.* 49: 854–866.

Sinclair, A. R. E., C. J. Krebs, J. N. M. Smith, and S. Boutin. 1988. Population biology of snowshoe hares. III. Nutrition, plant secondary compounds and food limitation. *J. Anim. Ecol.* 57: 787–806.

Sinclair, A. R. E., C. J. Krebs, R. Boonstra, S. Boutin, and R. Turkington. 2001. Testing hypotheses of community organization for the Kluane ecosystem. *In* C. J. Krebs, S. Boutin, and R. Boonstra (Eds.) *Ecosystem dynamics of the boreal forest: the Kluane project.* Oxford University Press, New York, pp. 407–496.

Smith, J. N. M., C. J. Krebs, A. R. E. Sinclair, and R. Boonstra. 1988. Population biology of snowshoe hares. II. Interactions with winter food plants. *J. Anim. Ecol.* 57: 269–286.

Sovell, J. R. 1993. *Attempt to determine the influence of parasitism on a snowshoe hare population during the peak and initial decline phases of a hare cycle.* M.Sc. thesis, University of Alberta, Edmonton, Alta.

Sovell, J. R. and J. C. Holmes. 1996. Efficacy of ivermectin against nematodes infecting field populations of snowshoe hares (*Lepus americanus*) in Yukon, Canada. *J. Wildl. Dis.* 32: 23–30.

Stefan, C. I. 1998. *Reproduction and pre-weaning juvenile survival in a cyclic population of snowshoe hares.* M.Sc. thesis, University of British Columbia, Vancouver, B.C.

Stenseth, N. C., W. Falck, K. Chan, O. N. Bjørnstad, M. O'Donoghue, H. Tong, R. Boonstra, S. Boutin, C. J. Krebs, and N. C. Yoccoz. 1998. From patterns to processes: phase and density dependencies in the Canadian lynx cycle. *Proc. Natl. Acad. Sci. USA* 95: 15,430–15,435.

Trostel, K., A. R. E. Sinclair, C. J. Walters, and C. J. Krebs. 1987. Can predation cause the 10-year hare cycle? *Oecologia* 74: 185–192.

Turkington, R., E. John, C. J. Krebs, M. R. T. Dale, V. O. Nams, R. Boonstra, S. Boutin, K. Martin, A. R. E. Sinclair, and J. N. M. Smith. 1998. The effects of NPK fertilization for nine years on boreal forest vegetation in northwestern Canada. *J. Veg. Sci.* 9: 333–346

Wolff, J. O. 1980. The role of habitat patchiness in the population dynamics of snowshoe hares. *Ecol. Monogr.* 50: 111–130.

Wolff, J. O. 1981. Refugia, dispersal, predation, and geographic variation in snowshoe hare cycles. *In* K. Myers and C. D. MacInnes (Eds.) *Proceedings of the World Lagomorph Conference,* University of Guelph, Guelph, Ont., pp. 441–449.

5

Evidence for Predator–Prey Cycles in a Bark Beetle

John D. Reeve and Peter Turchin

5.1 Introduction

The southern pine beetle, *Dendroctonus frontalis* Zimmermann (Coleoptera: Scolytidae), is an economically important pest of pine forests in the southern United States (Price et al. 1992). This native bark beetle is able to attack and kill living trees, typically loblolly (*Pinus taeda* L.) or shortleaf (*Pinus echinata* Mill.) pine, through a process of mass attack coordinated by pheromones emitted by the beetle (Payne 1980). During the attack process, thousands of beetles bore through the outer bark of the tree and begin constructing galleries in the phloem layer. Trees can respond to beetle attack by exuding resin from a network of ducts, but the large number of simultaneous attacks usually overcomes this defense, literally draining the resin from the tree. Oviposition and brood development then occur in the girdled (and ultimately dead) tree. Once a tree is fully colonized the attack process shifts to adjacent trees, often resulting in a cluster of freshly attacked trees, trees containing developing brood, and dead and vacated trees (Coulson 1980). These infestations can range in size from a single tree to tens of thousands, although the latter only occur in areas where no control methods are applied. Approximately six generations can be completed in a year in the southern United States (Ungerer et al. 1999).

Like many other forest insect pests, *D. frontalis* populations are characterized by a considerable degree of fluctuation. The longest time series available are Texas Forest Service records of infestations in southeast Texas since 1958 (figure 5.1a). These data suggest that the fluctuations have at least some periodic component, with major outbreaks occurring at intervals of 7–9 years

(1968, 1976, 1985, and 1992). A variety of different analyses, including standard time series analysis and response surface methodology (Turchin 1990, Turchin and Taylor 1992), suggest that *D. frontalis* dynamics are indeed cyclic and appear governed by some kind of delayed negative feedback acting on population growth (see chapter 1). This effect can be seen by plotting the realized per-capita rate of growth (*R*-values) over a year against population density in the previous year (figure 5.1b).

While these simple analyses can indicate the presence of potential regulating factors, delayed or otherwise, other approaches are needed to determine the underlying mechanisms. We review here the results of studies designed to evaluate the effect of natural enemies on *D. frontalis* dynamics, and to determine if they are a source of delayed negative feedback. The next section provides an overview of the impact and life cycle of *Thanasimus dubius* (F.) (Coleoptera: Cleridae), a prominent predator of *D. frontalis*. A simple model based on this predator–prey interaction is then used to test whether *T. dubius* can, in fact, generate delayed negative feedback and cycles in *D. frontalis*. This is followed by the results of a long-term exclusion experiment designed to evaluate the impact of natural enemies on *D. frontalis* at various times in the outbreak cycle. This experiment allowed us to determine if natural enemies have any impact on *D. frontalis* survival, and if so, whether this impact is constant through the cycle or if it produces delayed or immediate feedback effects. We also briefly examine the effect of competition on *D. frontalis* dynamics and conclude with a comparison of outbreak dynamics in this and other bark beetles.

5.2 Predator Impacts and Lifecycle

Like other bark beetles, *D. frontalis* is attacked by a variety of natural enemies, with each species arriving at a characteristic time in its lifecycle (Camors and Payne 1973, Dixon and Payne 1979b). The clerid predator *T. dubius* is a conspicuous and often very abundant component of this natural enemy community. Adult clerids are attracted by the pheromones emitted by *D. frontalis* (Vité and Williamson 1970) and arrive shortly after mass attack is initiated (Dixon and Payne 1979a). They catch and consume adult bark beetles as they arrive at the tree, and also mate and oviposit on the bark surface (Thatcher and Pickard 1966). After hatching, the larvae enter the phloem and prey on the egg, larval, and pupal stages of the developing bark beetle brood. Thus, all *D. frontalis* developmental stages are subject to attack by *T. dubius*, and both adult and larval predation must be evaluated in order to determine the total impact of this predator.

Predation by adult *T. dubius* can often be observed during mass attack in the field (Fiske 1908, Thatcher and Pickard 1966), but it has proven difficult to quantify under natural conditions. As a result, we turned to laboratory experiments that mimic these conditions as much as possible, but still allow an accurate estimation of predation rates (Reeve 1997). The experiments

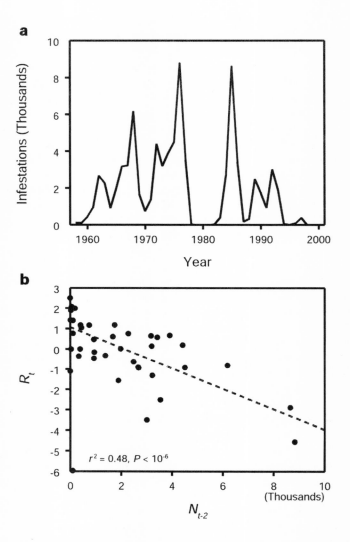

Figure 5.1 (a) Number of *D. frontalis* infestations in southeast Texas for the period 1958–99. (b) The per-capita rate of change, $R_t = \ln(N_t/N_{t-1})$, of *D. frontalis* infestations plotted on the number of infestations in the previous year, N_{t-2}, where N_t is the number of infestations at time t. The regression line and *P*-value exclude a single outlying point seen in the lower left-hand corner of the graph (with this point included, we have $r^2 = .27$, $P = .003$). The addition of N_{t-1} to the model produces no significant improvement in the fit. Data provided courtesy of the Texas Forest Service, Forest Pest Management, Lufkin, Texas.

made use of the fact that *D. frontalis* will readily attack freshly cut pine logs in the laboratory. We attached cages to these logs, and then introduced varying numbers of adult *T. dubius* and *D. frontalis* into the cages, using a range of densities spanning those observed in the field. Predator and prey were allowed to interact for 1 day, and then the proportion of *D. frontalis* successfully entering the log was determined by dissection. The results of this experiment indicate that adult clerids can significantly reduce the proportion of adult bark beetles successfully attacking the logs, from 60% to approximately 30% (figure 5.2a). Under field conditions, therefore, we expect *T. dubius* to slow (but probably not stop) the attack of *D. frontalis* on host trees and, ultimately, to reduce the overall number of trees attacked.

Analogous experiments with clerid larvae suggest that they can also reduce the survival of *D. frontalis* brood under field conditions. These experiments used another common bark beetle, *Ips grandicollis* (Eichh.), because this species can (unlike *D. frontalis*) complete its development in cut logs. Freshly cut logs were infested with different initial densities of adult *Ips*, seeded with a natural range of *T. dubius* eggs, and the number of bark beetle brood adults that finally emerged was recorded. A plot of the ratio of increase (the number of emerging female *Ips* per attacking female) against *T. dubius* eggs indicates that clerid larvae reduced the ratio of increase by 50% across all initial bark beetle densities (figure 5.2b), and it seems likely they would have a similar impact on *D. frontalis* reproduction. Comparable results have been found for clerid species attacking the brood of other bark beetles (Berryman 1970, Mills 1985, Weslien and Regnander 1992, Schroeder and Weslien 1994, Weslien 1994).

We have also examined the development of *T. dubius* in the field, to determine the number of predator generations per year compared with *D. frontalis*, and to uncover any time lags that could make it an agent of delayed negative feedback (Reeve et al. 1996, Reeve 2000). Emergence cages were attached to infested trees shortly after peak emergence of *D. frontalis*, and the pattern of *T. dubius* emergence observed over a period of 2 years. Similar to the results of previous laboratory studies (Nebeker and Purser 1980, Lawson and Morgan 1992), our data indicate that the lifecycle of *T. dubius* is initially well synchronized to its prey. Adult *T. dubius* arrive and oviposit at the beginning of *D. frontalis* attack, and the larvae appear to complete feeding just as the bark beetle brood emerge from the tree (Nebeker and Purser 1980, Lawson and Morgan 1992, Reeve et al. 1996, Reeve 2000). At this point, however, our field data indicate the larvae enter a prepupal stage of highly variable duration before emerging as adults. For a given tree, the emergence of adult clerids often occurs in several distinct episodes over a 2-year period, with most individuals emerging in spring and fall (figure 5.3). Individuals emerging after the first peak apparently enter a period of diapause before emerging in later peaks. Virtually no emergence occurs in the months of July and August and, in trees attacked in early summer, emergence is delayed until fall (Reeve 2000). Thus, *T. dubius* can have two generations per year, but with a significant fraction of the population having a much longer development

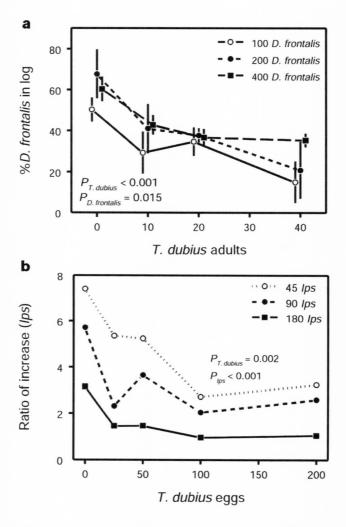

Figure 5.2 Impact of *T. dubius* on *D. frontalis* survival in behavioral experiments. (a) The percentage of adult *D. frontalis* successfully attacking a freshly cut log, as a function of adult *T. dubius* density and three initial densities (100, 200, or 400 adults) of *D. frontalis*. (b) Effect of predation by *T. dubius* larvae on the ratio of increase of *I. grandicollis*, a surrogate prey, as a function of the density of predator eggs and the initial densities of adult *I. grandicollis*. Here, *P*-values indicate the effect of predator or prey densities on attack success, or the ratio of increase using general linear models (see Reeve 1997 for details).

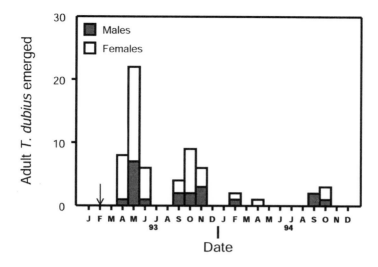

Figure 5.3 Emergence pattern of *T. dubius* from a tree attacked by *D. frontalis*. The arrow indicates the first date of sampling, which began shortly after *D. frontalis* emergence was complete.

time. This lifecycle would seem to make *T. dubius* a likely candidate for generating delayed negative feedback in *D. frontalis* dynamics.

We also examined the numerical response of *T. dubius* to fluctuations in *D. frontalis* density, using long-term survey data from the Kisatchie National Forest (central Louisiana) that encompassed one outbreak cycle. Multiple-funnel traps (Lindgren 1983) were baited with frontalin (the aggregation pheromone of *D. frontalis*) and pine turpentine, a combination attractive to both *D. frontalis* and *T. dubius* (Vité and Williamson 1970, Dixon and Payne 1980). The traps were deployed in four ranger districts, three to six traps per district, and serviced at 1–2-week intervals throughout the year. As a measure of population density in each district, we used the total number of *D. frontalis* and *T. dubius* caught per calendar year, divided by trapping effort (trap-days). Population densities for the entire Kisatchie are presented as means (±SE) across the four districts (figure 5.4).

The survey data showed a well-defined numerical response by *T. dubius* to increases in *D. frontalis* density, with both delayed and undelayed elements. At the beginning of the outbreak, predator densities lagged behind prey densities (1991–92) but eventually surpassed them in 1993, the year after peak *D. frontalis* densities. As the outbreak subsided to lower levels (1994–97), predator populations fluctuated in direct response to prey populations, but also remained at higher levels than before the outbreak. Predator densities were especially high the year before outbreak collapse (1997). When the outbreak finally collapsed (1998–2000), however, the predator population also returned to low but nonzero levels. It is likely that the *T. dubius* caught

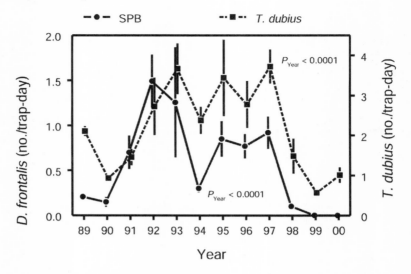

Figure 5.4 Mean trap catch (±SE) of *D. frontalis* and *T. dubius* across four ranger districts in the Kisatchie National Forest (central Louisiana). See text for further details.

in 1999–2000 are a combination of late-emerging adults (see figure 5.3) plus individuals developing on alternative prey, such as several common *Ips* bark beetles.

We then used multiple regression to examine the relationship between the per-capita growth rate of *D. frontalis* (R-values), and *D. frontalis* and *T. dubius* densities in the preceding years (see chapter 1 for methods). We regressed $R_t = \ln(N_t/N_{t-1})$ on N_{t-1} and P_{t-1} (a linear or Lotka–Volterra model), and on N_{t-1} and the predator–prey ratio P_{t-1}/N_{t-1} (a logistic ratio-dependent model), where N_t is *D. frontalis* and P_t is *T. dubius* density at time t (Berryman 1992, Berryman et al. 1995, Berryman and Gutierrez 1999). The logistic provided a much better fit than the linear model ($r^2 = .68$, $R_t = 5.58 - 4.02N_{t-1} - .59P_{t-1}/N_{t-1}$ versus $r^2 = .09$, $R_t = .64 - 1.08N_{t-1} - .76P_{t-1}$), suggesting that this predator–prey interaction is ratio-dependent, and providing further evidence that predators have a significant impact on *D. frontalis* populations. We were unable to include the year 2000 data in this analysis, because no *D. frontalis* were trapped that year. However, this apparent extinction of *D. frontalis* is consistent with the enormous predator–prey ratio seen in 1999 (>400 *T. dubius* per *D. frontalis* trapped), by far the largest ratio found in this time series.

Two other pieces of evidence also suggest that ratio dependence occurs in this system. First, laboratory behavioral studies demonstrated a pure ratio-dependent functional response in *T. dubius* (Reeve 1997). Second, an empirically derived forecasting method for *D. frontalis* associates high predator–prey ratios in trap catches with declining or low-level *D. frontalis* populations, and low ratios with increasing populations (Billings 1988). The key variables in this forecasting method are prey numbers and the predator–prey ratio, the same variables that were prominent in our survey data.

5.3 Predator–Prey Models

Our purpose for modeling the *T. dubius–D. frontalis* interaction was to determine if predation can generate cycles similar to those seen in nature (figure 5.1). A basic assumption of the model is that *T. dubius* is a major source of bark beetle mortality, which seems plausible given its potential impact on *D. frontalis* adults and larvae. The model uses simple functions to describe the interactions between predator and prey, but of necessity includes several details of the lifecycle of *T. dubius*. In particular, predator larvae are assumed to take twice as long to develop as *D. frontalis*, and once they complete development a certain proportion remain in diapause while others emerge as adults. The model also incorporates spring and fall bursts of emergence by *T. dubius* (see figure 5.3).

The model consists of three equations, one for *D. frontalis*, one for adult predators, and one for immature predators. For *D. frontalis*, we used a simple model of predation and intraspecific competition developed by Beddington et al. (1976):

$$N_{t+1} = N_t \exp[r_m(1 - N_t/K)] f(P_t, N_t), \tag{5.1}$$

where N_t and P_t are the densities of *D. frontalis* and adult *T. dubius* at time t. The function $f(P_t, N_t)$ is the fraction of prey surviving predation in each generation, while r_m is the maximum intrinsic rate of increase, and K is the carrying capacity for *D. frontalis*. We incorporate self-damping in the model because it is possible that competition, and perhaps a component of the natural enemy community, induce some immediate negative feedback in the system (see next section). The beetles are assumed to have six generations per year, in line with the predictions of temperature-driven developmental models (Ungerer et al. 1999). We used a discrete-time model, with a time step of one *D. frontalis* generation, because it simplifies the task of incorporating *T. dubius* development (see below). Because *D. frontalis* generations are overlapping in nature, it is an approximation of a continuous-time system.

The predator equations have the following structure:

$$Q_{t+1} = cN_t[1 - f(P_t, N_t)] + (1 - \varepsilon_t)Q_t, \tag{5.2}$$

$$P_{t+1} = (1 - \delta)P_t + \varepsilon_t Q_t, \tag{5.3}$$

where Q_t is the density of immature predators, ε_t is the proportion of immature predators that emerge as adults at time t, c is the conversion efficiency of *D. frontalis* to immature predators, P_t is the density of adult predators, and δ is the proportion of adult predators dying each *D. frontalis* generation. These equations incorporate the idea that immature predators enter a diapausing pool from which a fraction ε_t emerge at each time step, and that adult predators can persist for more than one *D. frontalis* generation. To model the twice-yearly emergence episodes that occur under field conditions (see figure 5.3), we let $\varepsilon_t = \varepsilon$ once every three time steps and zero otherwise.

We used two different functions for $f(P_t, N_t)$ the fraction of prey surviving predation. One was the simple Nicholson–Bailey (1935) model, for which $f(P_t, N_t) = \exp(-aP_t)$, with a defined as the search rate of the predator. We also employed a ratio-dependent model (Berryman and Gutierrez 1999), because there is evidence for ratio dependence in the dynamics of this system (see previous section). In this model $f(P_t, N_t) = \exp[-dP_t/(w + N_t)]$, where d represents the per-capita demand for resources by the predator. A number of different interpretations are possible for w, but for our purposes it indexes the amount of ratio dependence in the model; as $w \to 0$, the model reduces to the purely ratio-dependent search model developed by Thompson (1924). A similar model was developed by Getz and Mills (1996) and, with some redefining of parameters, the two models are identical. Both are also related to the type II function response model (Holling 1959).

Scaling the system leaves only four parameters in the Nicholson–Bailey version (r_m, ε, δ and the scaled parameter $a' = acK$), and five for the ratio-dependent model (r_m, ε, δ, $d' = dc$, and $w' = w/K$). We have reasonably accurate estimates of r_m, ε, and δ. Data from our long-term survey of *D. frontalis* density using pheromone-baited traps suggest $r_m \approx 1.8$ per year, or 0.3 per generation (see figure 5.4). Field data on *T. dubius* emergence imply $\varepsilon \approx 0.7$ (Reeve 2000), while studies of adult longevity suggest $\delta \approx 0.5$ (Lawson and Morgan 1992).

We then examined the behavior of the model through simulation, fixing r_m, ε, and δ at their estimated values while varying a', or d' and w'. The Nicholson–Bailey version of the model typically produced sustained cycles similar in period to the *D. frontalis* cycles observed in the field (figure 5.5a). This also occurred in the ratio-dependent version for moderate w' (figure 5.5b). As w' approached zero, however, the period of the cycles became longer, and the prey population alternated between long periods of extremely low and high density (figure 5.5c). This occurs because ratio-dependent predators are very efficient at low prey densities and can drive the prey close to extinction, increasing the time needed for prey populations to recover and hence the period of the cycles.

We also explored the effect of adding a small amount of *D. frontalis* immigration every generation, to mimic the flow of individuals from other populations or refuges. Very small amounts of immigration restored the cycles to periods similar to *D. frontalis* in the field (figure 5.5c). We conclude from this exercise that *T. dubius* can potentially induce cycles of the appropriate period in *D. frontalis* populations, regardless of the form of the predation model. If the interaction between *T. dubius* and *D. frontalis* approaches pure ratio dependence, however, our results suggest that addi-

Evidence for Predator–Prey Cycles in a Bark Beetle 101

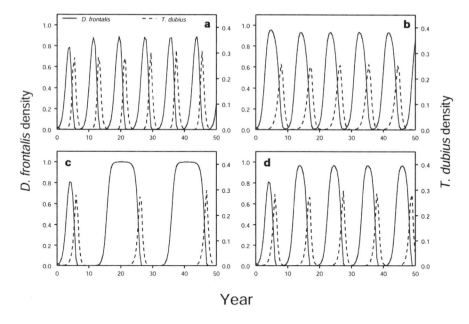

Figure 5.5 Simulations of predator–prey dynamics using a model patterned on the life histories of *D. frontalis* and *T. dubius* [equations (5.1)–(5.3)], using the parameter values $r_m = 0.3$, $\varepsilon = 0.7$, and $\delta = 0.5$ (see text). (a) Cycles with a period of approximately 8 years obtained using the Nicholson–Bailey model with $a' = 4$. (b) Cycles with a period of 9 years obtained using a ratio-dependent model with $d' = 2$ and $w' = 0.5$. (c) Cycles with a period approaching 20 years obtained for $d' = 2$ and $w' = 0.2$. (d) Cycles with a period of approximately 10 years obtained for $d' = 2$ and $w' = 0.2$ plus the immigration of 0.0001 prey per generation. Model trajectories were smoothed by averaging the densities for every three *D. frontalis* generations.

tional factors like immigration are necessary to produce the 7–9-year cycles observed in *D. frontalis*.

Models were also developed to examine two other possible causes for *D. frontalis* cycles: (1) the interaction between *D. frontalis* and its parasitoids, and (2) the interaction between host tree abundance and *D. frontalis*, with *D. frontalis* acting as a tree predator (Reeve and Turchin unpublished data). Neither model was able to generate cycles of the appropriate period. In particular, the *D. frontalis*–parasitoid model generally produced cycles with a shorter period than the natural system, while the periods in the *D. frontalis*–tree model were invariably too long. We conclude that the *D. frontalis*–*T. dubius* interaction is the most plausible explanation (among these models) for the cycles observed in *D. frontalis* populations.

5.4 Long-term Exclusion Study

A field experiment was designed to test the hypothesis that natural enemies are responsible for generating delayed negative feedback and population cycles in *D. frontalis* (Turchin et al. 1999). Stands of loblolly pine were located in the Kisatchie National Forest (central Louisiana), and a number of similarly sized mature trees were selected for the experiment. Cages were attached to the boles of experimental trees to protect areas of bark from natural enemies, while other trees were used as controls. All trees were then baited with the aggregation pheromone of *D. frontalis* (frontalin) to induce attack by natural beetle populations. As the trees came under attack, the caged areas were stocked with adult beetles, in a temporal pattern matching the natural attack process on exposed areas. Later, bark samples were removed to estimate the density of attacks and eggs and, eventually, logs were cut from the exposed and protected portions and placed in rearing cans to estimate the density of emerging adults (Turchin et al. 1999). Brood survival was estimated by dividing emerging adult density by egg density. The experiment was repeated for 5 years, covering one complete outbreak cycle.

The impact of natural enemies was evaluated by comparing *D. frontalis* brood survival in protected versus exposed areas (figure 5.6a). Survival of *D. frontalis* brood inside the cages did not differ significantly from that outside the cages during the phase of population increase (1990–91), indicating little impact of natural enemies at this time. A detectable difference in survival was observed during 1992, the peak of the bark beetle outbreak, but the largest difference occurred in 1993, during the first year of decline. Note that peak densities of *T. dubius* also occurred at this point (see figure 5.4). These results suggest that natural enemies (including *T. dubius*) act primarily as a delayed (second-order) feedback process, and may therefore play a role in causing *D. frontalis* cycles (although we cannot rule out other mechanisms that can generate delayed negative feedback, such as maternal effects, interactions with the host tree, or pathogens).

5.4.1. Competition

Although we have concentrated on the impact of predation, it is likely that competition also influences the dynamics of this system. Intraspecific competition within the host tree is known to have a major impact on *D. frontalis* reproduction and survival (Coulson et al. 1976, Reeve et al. 1998), as well as that of other bark beetles (Berryman and Pienaar 1973, Berryman 1974, Raffa and Berryman 1983, Anderbrant et al. 1985, Anderbrant 1990, Zhang et al. 1992). A number of reproductive variables are affected by high attack densities, including the number of eggs laid per attacking female, brood survival, and the ratio of increase (the number of emerging females per attacking female) (Reeve et al. 1998). In addition, *D. frontalis* brood compete for space in the phloem with cerambycid larvae and the bluestain fungus *Ophiostoma minus*, and it has been hypothesized that bluestain itself could

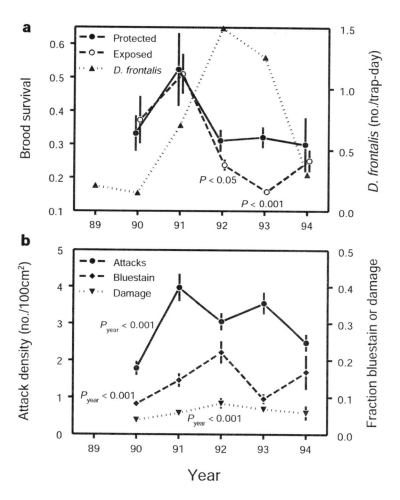

Figure 5.6 (a) Survival rates of *D. frontalis* brood during the long-term exclusion study in cage versus exposed areas (±1SE). Significant differences in survival rate occurred in 1992 and 1993. Also shown are average *D. frontalis* trap catches during the experiment (see also figure 5.4). (b) Attack density, and the fraction of phloem occupied by bluestain or damaged by cerambycids (±11SE) during the long-term exclusion study. For clarity, we plot attack density and bluestain levels averaged across both caged and exposed treatments because they were usually similar. Cerambycid levels were consistently lower in caged areas, and so here we plot the data only for the exposed areas (which better estimates their impact on the general *D. frontalis* population). Here, P-values indicate the effect of year in a two-way analysis of variance, treating year and cage as fixed effects.

drive *D. frontalis* cycles (Bridges 1985, Lombardero et al. 2000). Depending on how these competitive factors vary during the course of an outbreak, they could be another source of negative feedback in *D. frontalis* dynamics.

We evaluated the potential of competition to influence *D. frontalis* dynamics by examining the temporal pattern of attack density, and bluestain and cerambycid abundance, during the long-term exclusion study (Reeve and Turchin in preparation). All three competitive factors changed significantly through time (figure 5.6a), but none were at peak levels at the beginning of the collapse (1993), and so they would add little to the delayed effect of natural enemies. We note that all three factors were particularly low at the beginning of the outbreak and increased thereafter, and we speculate that competition may generate direct (undelayed) negative feedback in *D. frontalis* populations.

5.5 Conclusions

The studies reviewed here combine three different approaches for investigating the potential mechanisms explaining population cycles: (1) analysis of time series data, (2) models derived from the ecology and life histories of the organisms, and (3) field experiments designed to examine the mechanisms underlying these cycles (see also chapter 1). It is encouraging that these different approaches have produced a consistent picture for *D. frontalis* dynamics. Time series analysis indicated that the dynamics were governed by a second-order process, and our field experiments showed that natural enemies are the likely mechanism. The potential impact of *T. dubius* on *D. frontalis* survival, the time lags in its lifecycle, and the pattern of its numerical response suggest *T. dubius* is an important component of this second-order process. Intra- and interspecific competition may add negative feedback and stability to the system, but there was no evidence for a delay, and we would not expect these factors to contribute to *D. frontalis* cycles. Models incorporating *T. dubius* produced cycles similar to those observed in the field, further supporting the idea of predator–prey cycles. These cycles occurred in both the Nicholson–Bailey and the ratio-dependent versions of the model, and are likely determined in large part by the time lags inherent in this system. However, as the predator becomes more ratio-dependent the period of these cycles increased, but the addition of a small amount of immigration restored them to their original period. If the interaction between *T. dubius* and *D. frontalis* is truly ratio-dependent, this result suggests that two ingredients are necessary for cycles, the time delays inherent to the system and immigration (or some other factor) that sets a floor to prey densities. In this scenario, dispersal among populations would help maintain the 7–9-year cycles within each population, and also foster the global persistence of the system as a metapopulation.

Do natural enemies have a role in the cycles seen in other bark beetles, or is *D. frontalis* unique in this respect? In the western pine beetle, *Dendroctonus brevicomis*, the clerid beetle *Enoclerus lecontei* appears to be a major factor in the decline of outbreaks, in a similar way to *D. frontalis* (Berryman 1970). In both systems, outbreaks typically subside without significant depletion of suitable host trees, although there can be localized exceptions (Billings 1994, Berryman personal communication). There are also remarkable similarities in the lifecycles of these organisms. Both *D. frontalis* and *D. brevicomis* are multivoltine and each clerid species is bivoltine, but with some individuals taking a year or more to develop (Berryman 1970, Reeve 2000). At the other end of the spectrum, there are bark beetles whose natural enemies apparently have little influence on outbreak dynamics, such as mountain pine beetle (*Dendroctonus ponderosae*) or fir engraver (*Scolytus ventralis*) in North America, and perhaps *Ips typographus* in Eurasia. In these systems, outbreaks are apparently triggered by a decrease in host tree resistance (precipitated by age, drought, physical damage, defoliation, etc.), and are terminated when the supply of susceptible trees is exhausted (Berryman 1973, Raffa and Berryman 1983, Christiansen and Bakke 1988, Raffa 1988). Significant depletion of suitable host trees thus occurs during these outbreaks. However, there is some evidence that clerids have a significant impact on brood survival in *I. typographus*, and are particularly abundant during outbreak collapse (Mills 1985, Weslien and Regnander 1992, Weslien 1994), so it may represent an intermediate case.

The different types of outbreak dynamics seem to present a paradox, because the natural enemy communities are similar among these systems, and all include clerids or other related species. Why are the dynamics different in such similar communities? It may be no coincidence that the two species where predators seem important live in a relatively benign climate that permits multiple generations per year and continuous interaction between predator and prey. This could foster higher attack rates and a more effective numerical response by the predators, sufficient to reduce bark beetle numbers before depletion of suitable host trees. We would also expect to see some numerical response in systems like mountain pine beetle, *D. ponderosae*, but the response would be too weak or delayed to prevent destruction of the host tree population. Studies similar to our long-term exclusion experiment, as well as long-term monitoring of bark beetle and natural enemy populations, would seem necessary to evaluate the role of natural enemies in these other bark beetle species. If *D. frontalis* is any indication, it is critically important to replicate the experiment throughout the entire outbreak cycle because, in our experiments, significant effects were only detected during the decline phase.

ACKNOWLEDGMENTS

We thank Alan Berryman and Kim Cuddington for their helpful comments on the manuscript, Brian L. Strom for many discussions on bark beetle dynamics, Jonny S. Fryar, Douglas J. Rhodes, and John A. Simpson for vital technical assistance, and the Kisatchie National Forest for providing access to field sites. Financial support was provided by the Southern Research Station, USDA Forest Service, and the National Science Foundation (DEB 9509237).

REFERENCES

Anderbrant, O. 1990. Gallery construction and oviposition of the bark beetle *Ips typographus* (Coleoptera: Scolytidae) at different breeding densities. *Ecol. Entomol.* 15: 3–8.

Anderbrant, O., F. Schlyter, and G. Birgersson. 1985. Intraspecific competition affecting parents and offspring in the bark beetle *Ips typographus*. *Oikos* 45: 89–98.

Beddington, J. R., C. A. Free, and J. H. Lawton. 1976. Concepts of stability and resilience in predator–prey models. *J. Anim. Ecol.* 45: 791–816.

Berryman, A. A. 1970. Evaluation of insect predators of the western pine beetle. In R. W. Stark and D. L. Dahlsten (Eds.) *Studies on the population dynamics of the western pine beetle*, Dendroctonus brevicomis LeConte *(Coleoptera: Scolytidae)*. University of California, Berkeley, Division of Agricultural Sciences, pp. 102–112.

Berryman, A. A. 1973. Population dynamics of the fir engraver, *Scolytus ventralis* (Coleoptera: Scolytidae). I. Analysis of population behavior and survival from 1964 to 1971. *Can. Entomol.* 105: 1465–1488.

Berryman, A. A. 1974. Dynamics of bark beetle populations: towards a general productivity model. *Environ. Entomol.* 3: 579–585.

Berryman, A. A. 1992. The origins and evolution of predator–prey theory. *Ecology* 73: 1530–1535.

Berryman, A. A. and A. P. Gutierrez. 1999. Dynamics of insect predator–prey interactions. In C. B. Huffaker and A. P. Gutierrez (Eds.) *Ecological entomology* (2nd edition). John Wiley, New York, pp. 389–423.

Berryman, A. A. and L. V. Pienaar. 1973. Simulation of intraspecific competition and survival of *Scolytus ventralis* broods (Coleoptera: Scolytidae). *Environ. Entomol.* 2: 447–459.

Berryman, A. A., A. P. Gutierrez, and R. Arditi. 1995. Credible, parsimonious and useful predator–prey models—a reply to Abrams, Gleeson, and Sarnelle. *Ecology* 76: 1980–1985.

Billings, R. F. 1988. Forecasting southern pine beetle infestation trends with pheromone traps. In T. L. Payne and H. Sarenmaa (Eds.) *Integrated control of scolytid bark beetles*. Virginia Polytechnic Institute and State University, Blacksburg, Va., pp. 295–306.

Billings, R. F. 1994. Southern pine beetle: impact on wilderness and non-wilderness areas. *Texas For.* 35: 16–17.

Bridges, J. R. 1985. Relationships of symbiotic fungi to southern pine beetle population trends. In S. J. Branham and R. C. Thatcher (Eds.) *Integrated pest management research symposium: the proceedings*. USDA Forest Service, Southern Forest Experiment Station, New Orleans, La., pp. 127–135.

Camors, F. B. and T. L. Payne. 1973. Sequence of arrival of entomophagous insects to trees infested with southern pine beetle. *Environ. Entomol.* 2: 267–270.

Christiansen, E. and A. Bakke. 1988. The spruce bark beetle of Eurasia. In A. A. Berryman (Ed.) *Dynamics of forest insect populations: patterns, causes, implications*. Plenum Press, New York, pp. 479–503.

Coulson, R. N. 1980. Population dynamics. In R. C. Thatcher, J. L. Searcy, J. E. Coster, and G. D. Hertel (Eds.) *The southern pine beetle*. USDA Forest Service Technical Bulletin 1631, pp. 70–105.

Coulson, R. N., A. M. Mayyasi, J. L. Foltz, F. P. Hain, and W. C. Martin. 1976. Resource utilization by the southern pine beetle, *Dendroctonus frontalis* (Coleoptera: Scolytidae). *Can. Entomol.* 108: 353–362.

Dixon, W. N. and T. L. Payne. 1979a. Aggregation of *Thanasimus dubius* on trees under mass-attack by the southern pine beetle. *Environ. Entomol.* 8: 178–181.

Dixon, W. N. and T. L. Payne. 1979b. *Sequence of arrival and spatial distribution of entomophagous and associate insects on southern pine beetle-infested trees*. Texas Agricultural Experiment Station, Miscellaneous Publication 1432.

Dixon, W. N. and T. L. Payne. 1980. Attraction of entomophagous and associate insects of the southern pine beetle to beetle- and host tree-produced volatiles. *J. Georgia Entomol. Soc.* 15: 378–389.

Fiske, W. F. 1908. Notes on insect enemies of wood boring Coleoptera. *Proc. Entomol. Soc. Wash.* 10: 23–27.

Getz, W. M. and N. J. Mills. 1996. Host-parasitoid coexistence and egg-limited encounter rates. *Am. Nat.* 148: 333–347.

Holling, C. S. 1959. Some characteristics of simple types of predation and parasitism. *Can. Entomol.* 91: 385–398.

Lawson, S. A. and F. D. Morgan. 1992. Rearing of two predators, *Thanasimus dubius* and *Temnochila virescens*, for the biological control of *Ips grandicollis* in Australia. *Entomol. Exp. Appl.* 65: 225–233.

Lindgren, B. S. 1983. A multiple funnel trap for scolytid beetles (Coleoptera). *Can. Entomol.* 115: 299–302.

Lombardero, M. J., K. D. Klepzig, J. C. Moser, and M. P. Ayres. 2000. Biology, demography and community interactions of *Tarsonemus* (Acarina: Tarsonemidae) mites phoretic on *Dendroctonus frontalis* (Coleoptera: Scolytidae). *Agric. For. Entomol.* 2: 193–202.

Mills, N. J. 1985. Some observations on the role of predation in the natural regulation of *Ips typographus* populations. *Z. Angew. Entomol.* 99: 209–320.

Nebeker, T. E. and G. C. Purser. 1980. Relationship of temperature and prey type to development time of the bark beetle predator *Thanasimus dubius* (Coleoptera: Scolytidae). *Can. Entomol.* 112: 179–184.

Nicholson, A. J. and V. A. Bailey. 1935. The balance of animal populations. *Proc. Zool. Soc., Lond.* 3: 551–598.

Payne, T. L. 1980. Life history and habits. In R. C. Thatcher, J. L. Searcy, J. E. Coster, and G. D. Hertel (Eds.) *The southern pine beetle*. USDA Forest Service Technical Bulletin 1631, pp. 7–28.

Price, T. S., C. Doggett, J. M. Pye, and T. P. Holmes. 1992. *A history of southern pine beetle outbreaks in the southeastern United States*. Georgia Forestry Commission, Macon, Ga.

Raffa, K. F. 1988. The mountain pine beetle in western North America. In A. A. Berryman (Ed.) *Dynamics of forest insect populations: patterns, causes, implications*. Plenum Press, New York, pp. 505–530.

Raffa, K. F. and A. A. Berryman. 1983. The role of host plant resistance in the colonization behavior and ecology of bark beetles (Coleoptera: Scolytidae). *Ecol. Monogr.* 53: 27–49.

Reeve, J. D. 1997. Predation and bark beetle dynamics. *Oecologia* 112: 48–54.

Reeve, J. D. 2000. Delayed emergence in a bark beetle predator: implications for population dynamics. *Agric. For. Entomol.* 2: 233–240.

Reeve, J. D., J. A. Simpson, and J. S. Fryar. 1996. Extended development in *Thanasimus dubius* (F.) (Coleoptera: Cleridae), a predator of the southern pine beetle. *J. Entomol. Sci.* 31: 123–131.

Reeve, J. D., D. J. Rhodes, and P. Turchin. 1998. Scramble competition in the southern pine beetle. *Ecol. Entomol.* 23: 433–443.

Schroeder, L. M. and J. Weslien. 1994. Interactions between the phloem-feeding species *Tomicus piniperda* (Coleoptera: Scolytidae) and *Acanthocinus aedilis* (Coleoptera: Cerambycidae), and the predator *Thanasimus formicarius* (Coleoptera: Cleridae) with special reference to brood production. *Entomophaga* 39: 149–157.

Thatcher, R. C. and L. S. Pickard. 1966. The clerid beetle, *Thanasimus dubius*, as a predator of the southern pine beetle. *J. Econ. Entomol.* 59: 955–957.

Thompson, W. R. 1924. La théorie mathématique de l'action des parasites entomophages et le facteur du hasard. *Ann. Fac. Soc. Marseille* 2: 69–89.

Turchin, P. 1990. Rarity of density dependence or population regulation with lags? *Nature* 344: 660–663.

Turchin, P. and A. D. Taylor. 1992. Complex dynamics in ecological time series. *Ecology* 73: 289–305.

Turchin, P., A. D. Taylor, and J. D. Reeve. 1999. Dynamical role of predators in population cycles of a forest insect: an experimental test. *Science* 285: 1068–1071.

Ungerer, M. J., M. P. Ayres, and M. J. Lombardero. 1999. Climate and the northern distribution limits of *Dendroctonus frontalis* Zimmermann (Coleoptera: Scolytidae). *J. Biogeogr.* 26: 1133–1145.

Vité, J. P. and D. L. Williamson. 1970. *Thanasimus dubius*: prey perception. *J. Insect Physiol.* 106: 233–239.

Weslien, J. 1994. Interaction within and between species at different densities of the bark beetle *Ips typographus* and its predator *Thanasimus formicarius*. *Entomol. Exp. Appl.* 71: 133–143.

Weslien, J. and J. Regnander. 1992. The influence of natural enemies on brood production in *Ips typographus* (Coleoptera: Scolytidae) with special reference to egg-laying and predation by *Thanasimus formicarius* (L.) (Coleoptera: Cleridae). *Entomophaga* 37: 333–342.

Zhang, Q. H., J. A. Byers, and F. Schlyter. 1992. Optimal attack density in the larch bark beetle, *Ips cembrae* (Coleoptera: Scolytidae). *J. Appl. Ecol.* 29: 672–678.

6

Parasitic Worms and Population Cycles of Red Grouse

Peter J. Hudson, Andrew P. Dobson, and David Newborn

6.1 Historical Introduction to Grouse Research

Many years before Charles Elton collected the detailed data on fur returns to The Hudson's Bay Trading Company, or described the regular fluctuations in small mammal numbers, scientists and naturalists had observed and were proposing explanations for the cause of periodic crashes in numbers of red grouse known as "grouse disease." MacDonald (1883) claimed "that it was more than eighty years since the alarm of grouse disease was sounded in this country," implying that naturalists were starting to examine the phenomenon nearly 200 years ago. In 1873, The House of Commons established a Select Committee to consider the game laws of the United Kingdom and, since this had followed a year of particularly severe population collapse in red grouse numbers, they took exhaustive evidence on a wide range of possible causes of "grouse disease." An examination of the letters in *The Times* and *The Field* shows that the debate over the cause of the population crashes was contentious and as heated as many of the recent debates over the causes of population cycles.

Scientific studies were initiated by Cobbold (1873) who examined grouse killed during a population crash, published a pamphlet that described the presence of large numbers of "strongle worms," and advocated the theory that the cause of grouse disease was wholly due to the presence of nematode worms. In 1905, the Board of Agriculture appointed a Committee of Inquiry on Grouse Disease to investigate the life history of the parasite and the causes of "grouse disease." The extensive survey and detailed analysis was quite remarkable for the time, and was presented in a two-volume publication

(Lovat 1911). The Committee surveyed grouse populations, undertook experiments and, after nearly 2000 dissections, came to the conclusion that "the strongyle worm, and the strongyle worm alone, is the immediate *causa causans* of *adult* 'Grouse Disease.'" The Principal Field Officer was E. A. Wilson, a gifted artist and scientist who was later appointed as the Scientific Director to Captain Scott's Antarctic expedition on the *Terra Nova*. Unfortunately, Wilson never saw the production of the final report as he died with Scott during their return from the South Pole.

Mackenzie (1952) presented and described the bag records of grouse species shot in the United Kingdom, and Moran (1952) undertook time series, analysis that described the tendency of the populations to show regular oscillations. Lack (1954) was concerned that the oscillations were not as regular as the Canadian lynx time series, and postulated that the main cause of the cycles was an interaction between parasites and food. He based this idea not on new data, but on a reading of the main report by the Inquiry on Grouse Disease (Lovat 1911). The red grouse remained a focus of population research and, in 1956, Professor Wynne-Edwards established a unit of Aberdeen University to "test the Animal Dispersion Hypothesis." Later, as his ideas on group selection were generally dismissed, workers focused on the role of spacing behavior in the population dynamics of red grouse. After 30 years of detailed work, Wynne-Edwards (1986) summarized his perspective on the findings thus:

> Grouse disease as a lethal epidemic was a misunderstanding: instead, as predicted at the outset, it's the birds themselves that by mutual competition determine their own population density.... The successful ones occupy the habitat in a virtually continuous mosaic. The surplus ones become outcasts, and expendable; all of them are normally dead from secondary causes before the spring when breeding starts. 'Grouse disease' was a mistaken diagnosis of the after effects of social exclusion.

This statement is in direct contrast with the clear statement given by Lovat (1911), and underemphasizes the detailed experiments and monitoring that has been undertaken on the role of spacing behavior by the workers in northeast Scotland. Since this book is concerned with the hypothesis that grouse cycles are caused by trophic interactions, we will not review the large literature on the possible role of spacing behavior in grouse population dynamics (see, e.g., Moss and Watson 1991).

By the 1970s it was generally accepted amongst population biologists that spacing behavior determined grouse density and that grouse disease was a consequence of the behavior of the birds. This idea fitted the belief portrayed in text books that in "natural systems" parasites were selected to be benign to their hosts and not to have an important long-term impact on host dynamics (e.g., Ricklefs 1979). However, Anderson and May (1978) synthesized the disciplines of population biology and parasitology to provide a sound mathematical framework that captured the dynamics of the parasite–host relationship. They specifically identified the destabilizing characteristics of the parasite–host relationship that would generate oscillations in host abundance. Their models showed that a random or regular distribution of parasites in the host population, parasite-induced reduction in host fecundity, and time delays in parasite development would all destabilize a host population, although the final dynamics would be a consequence of the tension between these processes and other stabilizing mechanisms. Interestingly, their work highlighted the importance of parasite-induced reduction in host fecundity and emphasized that, depending on the relative degree of parasite aggregation within the host population, this had to be larger than the relative impacts on survival. This observation is interesting in the context of the previous studies on "grouse disease," where workers considered only the lethal effects of parasites and not the sublethal effects that provided a more focused explanation of red grouse cycles.

From this historical perspective the question arises "What are the sublethal effects of parasites on red grouse and what role do they play in the population dynamics of red grouse?" The social behavior hypothesis proposes that parasites are a consequence of the effects of social exclusion (Jenkins et al. 1963). In contrast, the parasite hypothesis suggests that parasites reduce fecundity of grouse in a way that causes the grouse population to cycle in abundance. This chapter examines this hypothesis in more detail and summarizes a series of published studies (Hudson et al. 1985, 1992a,b, 1999; Hudson 1986a,b, 1992; Dobson and Hudson 1992, 1995; Hudson and Dobson 1996).

6.2 Time Series Analysis

Red grouse population time series occur in the form of bag records and sample counts within populations. Each privately owned grouse moor seeks to maximize its long-term hunting returns which, in reality, means harvesting to a level where a suitable breeding stock is left for the following year. Hunting is not regulated, and the individual landowner decides how many birds can be harvested. For more than 150 years, landowners have kept detailed and accurate counts of numbers harvested each year on different managed estates. Since each is managed independently and often separated from others by natural watersheds, workers have tended to consider each time series as independent. Population counts show that bag records are usually a fair reflection of abundance (figure 6.1), although since hunting takes place in the fall, approximately two thirds of the harvest is of immature birds and so breeding production influences bag records. Overall, the variance in hunting mortality is greater at population densities less than 100 birds/km^2, probably because grouse are aggregated and counts less accurate at low densities (Hudson et al. 1999).

Time series analysis of the bag records shows that the majority of time series (Potts et al. 1984, Hudson et al. 1985, Hudson 1992) show significant autocorrelation coefficients at half the cycle period, which is typical of phase-forgetting quasi-cycles (Gurney and Nisbett 1998). Significant partial

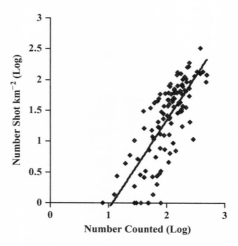

Figure 6.1 Relationship between grouse harvest and population density per square kilometer. While there is evidence that hunting records at a large scale reflect density estimates at a lower scale, the variance in the counts is greater at densities less than 100 birds/km^2. This is because at low grouse densities grouse are aggregated and the count areas do not reflect this aggregation well (Hudson et al. 1999).

autocorrelation coefficients indicate a second-order density dependence, and spectral analysis shows that most populations produce cycles with a period of between 4 and 6 years (figure 6.2). Generally, the cycle period increases with latitude, with longer cycles in Scotland than in England, and this is apparently associated with lower breeding production in the north. Not all populations produce cyclic bag records. Those from small areas of moorland and on the drier, eastern side of the country have a lesser tendency to oscillate (Hudson et al. 1985, Hudson 1992). This is important, since any explanation for the cause of cycles must account for why some populations are cyclic and others are not, and must also explain variations in cycle period.

6.3 Nematode Parasites

Originally described by Cobold (1873) as *Trichostrongylus pergracilis*, the grouse cecal nematode was later included with *Trichostrongylus tenuis* (Mehlin 1846). The nematodes inhabit the large blind-ending ceca of red grouse, where they cause internal inflammation and bleeding (Watson et al. 1987). The parasite has a direct lifecycle with no intermediate hosts. Adults produce eggs at a constant rate with no evidence of density-dependent fecundity (Hudson and Dobson 1996). The first two larval stages stay within the feces, and the third stage infective larvae retain the sheath of

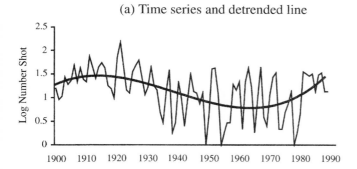
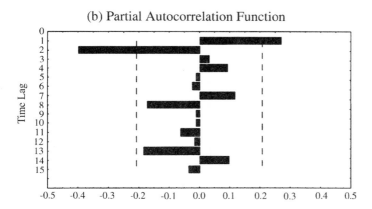
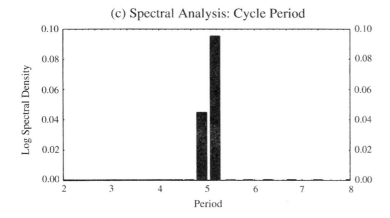

Figure 6.2 Time series analysis of grouse harvesting records. (a) Numbers shot per annum is nonstationary, so the time series was detrended using a third-order polynomial, and the residuals used for the analysis. (b) Partial autocorrelogram with Bartlett lines indicating the level of significance. (c) Smoothed spectral analysis of time series indicating a cycle period of approximately 5 years.

the second larval stage, leave the cecal pat, and ascend the growing shoots of the principal food plant of red grouse, heather *Calluna vulgaris* (Saunders et al. 1999, 2000). The larval stage requires moist conditions for development and to aid mobility up the plant (Hudson 1986b, Saunders et al. 1999), and infection occurs when the grouse eat the growing tips. Uptake of worms by immature grouse occurs principally during July and August, when temperature allows development (Hudson 1996, Hudson and Dobson 1996). Some infective stages may shed the coat of the second larval stage and remain dormant in the gut mucosa before commencing development in late winter (Shaw 1988). The parasites show an aggregated distribution within the adult host population (Hudson et al. 1992a), although the degree of aggregation is weak (k of the negative binomial $= 1$) compared with the majority of parasitic infections (Shaw and Dobson 1995). This is probably a consequence of the low level of acquired resistance within the host population (Hudson and Dobson 1996).

To relate changes in the population to changes in parasite burdens, red grouse numbers have been carefully monitored annually since 1979 on 1 km^2 plots. Sample counts are taken every spring to record breeding density and in July to record number of chicks reared per female. The per-capita rate of change of the population decreases, and breeding mortality increases, as the intensity of worm infection increases (figure 6.3). Replicated field experiments have shown that grouse with experimentally reduced infections have higher brood production and survival (Hudson 1986b, Hudson et al. 1992a). This finding shows that parasitic infections reduce grouse breeding production, and is in line with the model of May and Anderson (1978), which predicts that parasite-induced reduction in host fecundity is destabilizing.

6.4 Population Model

Dobson and Hudson (1992) explored the dynamics of the interaction between red grouse and *T. tenuis* with Anderson–May models. We incorporated a third equation to describe changes in the size of free-living parasites and included a description of arrested larval development. The model is based on the lifecycle of the parasite and includes the parasite-induced impacts on survival and brood production of the grouse, as determined by field experiments. Simulations with the model suggest that time delays caused by parasite-arrested development (or hypobiosis) are not the main cause of population oscillations. Cyclic oscillations in the model occur when the ratio of parasite-induced reduction in host fecundity to parasite-induced reduction in host survival is greater than the degree of parasite aggregation within the host population. In other words, the relative impact on fecundity destabilizes the host population and is the principal cause of the oscillation. The reproductive rate of the grouse and the life expectancy of the free-living stages determine the cycle period (figure 6.4).

Parasitic Worms and Population Cycles of Red Grouse   115

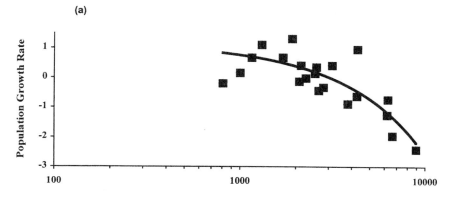

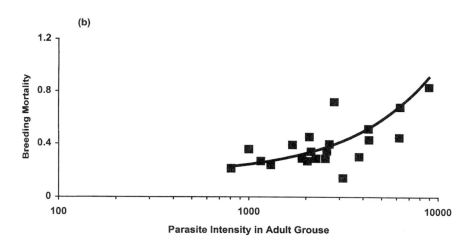

Figure 6.3 The relationship between geometric mean parasite intensities and population change. (a) Data from a 22-year longitudinal study of grouse populations. (b) Breeding mortality estimated as killing mortality (log difference, numbers of chicks at 6 weeks minus maximum clutch size taken as 12) in relation to geometric mean intensity of parasite infection in adult grouse.

The role of arrested development is interesting since this effectively adds a time delay between the production of the infective stage and the impact of the parasite, and it is well known that time delays can be destabilizing (May 1976). In this system, the time delay increases the cycle period but tends to dampen the oscillations. This occurs because the arrested larvae have no impact on their host until they emerge, but die when the host dies, effectively increasing the mortality of the free-living stages and dampening the oscillations (Dobson and Hudson 1992).

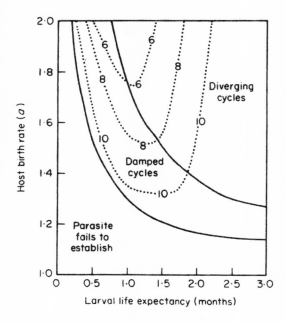

Figure 6.4 The influence of larval longevity and host fecundity on the dynamics of red grouse when there is no arrestment in the model of Dobson and Hudson (1992).

6.5 Experiments

Our simulation of the grouse–nematode interaction predicts that red grouse populations will cycle with a period of 5 to 10 years, and with an amplitude and pattern (slow up and fast down) similar to the observed time series. This suggests that the interaction with parasitic worms is a plausible hypothesis for the grouse cycle. The problem is, how to test this hypothesis to the exclusion of others. The model provides a possible way to do this: First, we included in the model a description of worm mortality caused by the application of the anthelmintic Levamisole hydrocholoride (Hudson et al. 1999). Second, we used the model to predict the level of treatment needed to suppress the large-amplitude cycles normally observed (figure 6.5a). Notice that we would need to treat more than 20% of the grouse population to have a large impact on grouse cycles, so our experiments were designed to treat 20% or more of the grouse population. It is interesting to note that removal of a small proportion of worms has a large influence on host dynamics, but worm eradication would require the treatment of nearly all grouse. The model identifies an interesting compensation similar to the paradox of enrichment. Treatment of the grouse reduces the size of the worm population but increases grouse fecundity, such that the grouse population rises and there are more hosts for the surviving parasites to infect. Consequently, the more grouse that

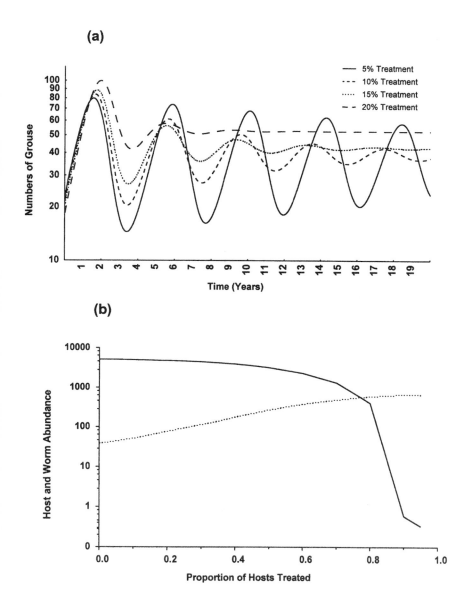

Figure 6.5 (a) Predicted decreases in cycle amplitude of grouse populations with increasing proportion of anthelmintic-treated grouse. (b) Equilibrium grouse and worm populations obtained by increasing levels of treatment with anthelmintic. Treatment results in higher grouse fecundity and, consequently, an increase in transmission such that more than 96% of the population needs to be treated to eradicate the worms (after Hudson et al. 1999).

are treated the greater the force of infection on the untreated birds and their offspring. Ultimately, eradication of the parasite requires that more than 90% of the grouse be treated (figure 6.5b).

6.5.1 Experimental Methods

Six study areas approximately $20\,km^2$ in size were located in either North Yorkshire or County Durham in northern England in independently managed grouse populations. Using the model with previous estimates of grouse abundance and worm burdens, we predicted that the first crash would occur in 1989. During the winter before the crash large numbers of grouse were caught at night by dazzling the birds (Hudson and Newborn 1995). On four of the six study areas, grouse were treated orally with the anthelmintic Levamisole hydrochloride to reduce intensities of worm infection, and marked with a patagial wing flash. The remaining two populations were left untreated. On three of the treated populations, about 20% of the population were caught, treated, and tagged, but on the third only about 15% were treated. The next population crash was predicted in 1993, and in this year two of the previously treated areas were retreated. Bag records from managed estates were used to monitor changes in the grouse populations. Grouse bag records were used as a proxy for grouse abundance because resources were not available to monitor all sites in detail. The relationship between density estimated on $1\,km^2$ areas and numbers harvested at a much larger scale ($20\,km^2$) is reasonably good (see figure 6.1).

6.5.2 Results

The overall effect of treating grouse with an anthelmintic was to reduce the collapse of the population from its peak (figure 6.6). Note that the amplitude of the cycle was suppressed on all treated plots, as predicted by the model (figure 6.5a); fewer grouse were harvested from the site where a lower proportion of grouse were treated, but this was expected from the model. In general, the removal of parasites reduced the amplitude of the oscillations implying that, on these sites and at these times, parasitism was responsible for the observed population fluctuations. Of course, our results do not prove that other factors could not be involved, on other sites or at other times, but leaves little doubt that parasites were the principal cause of population cycles on our study sites. We are now testing the hypothesis on other populations where parasites are not thought to be involved in the population cycles.

6.5.3 Criticisms

Proponents of the spacing hypothesis have criticized our experiment, particularly our contention that "these results show parasites were both sufficient and necessary in causing cycles in these populations" (chap. 9, Lambin et al.

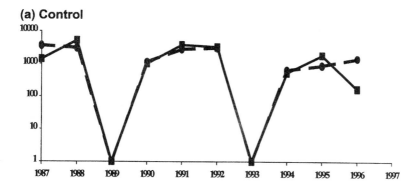
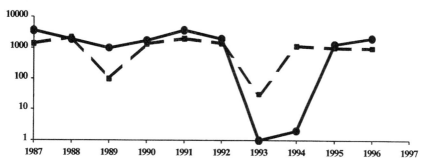
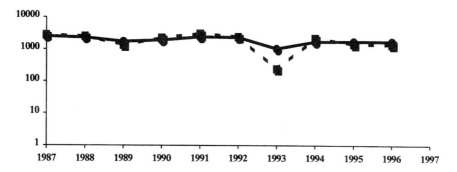

Figure 6.6 Experimental treatment of grouse with anthelmintic just prior to the population crash reduces the severity of the crash (numbers of grouse shot are expressed in logarithms of numbers + 1). The experiment was conducted on four treated populations, two treated once in 1989 and two treated twice, once in 1989 and once in 1993 (after Hudson et al. 1998).

1999; but see Hudson et al. 1999). To some degree this is an argument over the semantics of the terms "sufficient and necessary," but there are two specific criticisms. First, bag records were used in the analysis and these may overestimate the variance in population size because hunters tend not to harvest at low densities. Thus, during our experiment, the control populations were not harvested when the populations were very low. A superficial examination of the data (figure 6.1) confirms that the variance does in fact increase when grouse densities are below 100 birds/km^2. However, a more detailed analysis of the count data from our main study area shows that the variance falls with the mean, such that at low densities there was more variation between count areas than recorded at high densities. In this respect the variance is really in the count data rather than the bag records since, when grouse numbers fall, the population becomes aggregated in certain areas and is not so evenly distributed throughout the available habitat. As such, bag records may well provide a better reflection of what is happening at the population level than the more precise counts conducted over a small area. Of course, a full population count across the whole area would have been the best answer, but was clearly impractical.

The second criticism is that the oscillations are only reduced, and a residual oscillation remains that could be caused by some other mechanism. Again, this is somewhat semantic because the residual oscillation cannot be detected statistically. Either way, our model predicts a residual oscillation even in populations where 20% of the birds are treated (see figure 6.5a). In other words, we expected the oscillations to be reduced not eliminated, and so the experimental results are in complete agreement with the hypothesized effect.

A third criticism is that the treatment would have killed the whole community of gut helminths and not just *T. tenuis*. Of course this is true. However, analysis of birds at the time of treatment indicated that other parasites were rarely present and, when they were, their numbers were low. Tapeworms, for example, are relatively rare in spring compared with autumn. Thus, although all gut helminths would have been killed by the treatment, the dominant parasite is without doubt *T. tenuis*. Even if there were synergistic interactions between parasites, this would tend to support rather than refute the parasite hypothesis.

In reviewing this paper Robert Moss raised another criticism, that the experimental results could also be explained by the kin-selection/spacing-behavior hypothesis. He argued that this hypothesis would predict more young per adult in the summer following treatment, which would be followed by enhanced recruitment the next spring, resulting in larger kin clusters and, thereby, enhanced recruitment in the following spring as well. Thus, he feels that our experiment cannot distinguish between the predictions of the parasite hypothesis and the kin-selection hypothesis. However, we feel that this scenario is irrelevant, since the parasites are the actual cause of these population changes. Interestingly, this interpretation was part of the original kin-selection hypothesis proposed by Hudson et al. (1985).

6.6 Noncyclic Populations

In many species, including red grouse, the tendency for populations to cycle varies from place to place and, indeed, from time to time. Understanding the differences between cyclic and noncyclic populations by comparative study provides a natural way of exploring alternative hypotheses. Some grouse populations, particularly those inhabiting small moors and the eastern side of the United Kingdom, where rainfall is low, are less likely to exhibit cyclic hunting records (Hudson et al. 1985, Hudson 1992). These drier eastern moors tend to be freely drained heath dominated by pure stands of heather, with little *Sphagnum* moss and, consequently, a shallow peat layer. In contrast, moors in wetter areas have a mixture of heather and grasses, a thick layer of *Sphagnum* moss, and a deep peat layer that may be up to several meters thick.

The free-living stages of trichostrongyle worms require moisture for development to the third-stage infective larvae. A film of moisture is also required to aid worm movement up the vegetation to where it can be ingested by its host (Saunders et al. 2000). As moisture levels decline, fewer infective stages are available to infect grouse (Saunders et al. 1999, 2000). Hudson et al. (1985) recorded significantly lower worm intensities from non-cyclic populations, where only 4% of the birds had harmful infections compared with 24% from cyclic populations. As there was no difference in the densities of birds in the two regions, we concluded that the differences in infection levels probably reflected differences in survival of free-living nematodes and the exposure of birds to the infective stages. Thus, the absence of grouse cycles on many freely drained heather moors is explained by, and provides support for, the parasite-interaction hypothesis.

It is interesting that grouse populations in parts of Scotland have shown a reduced tendency to oscillate in recent years (Hudson 1992), a tendency that does not appear to be associated with large-scale changes in habitat. The explanation of this phenomenon requires consideration of other natural enemies, particularly birds of prey and the louping ill virus (see below).

6.7 Louping Ill Virus

Red grouse are killed by a variety of natural enemies besides the strongyle worm, and some of these may have a profound impact on the dynamics of grouse and nematode populations. The louping ill virus is transmitted between sheep and grouse by the sheep tick, *Ixodes ricinus*. The virus can cause 80% mortality in exposed grouse, whereas mortality in sheep is variable and depends on breed and history of exposure (Reid et al. 1978). Only sheep and red grouse are viraemic hosts, and permit the amplification of the virus within the host to a concentration that leads to subsequent infection of other uninfected ticks. In most moorland areas where sheep suffer from the virus, shepherds vaccinate their sheep and treat them with acaricides so, in reality, sheep play no role in the amplification of the virus. Grouse alone are unable to maintain the virus so a large mammalian host is required for female ticks to complete their lifecycle. As no vertical transmission occurs between female ticks and their eggs, the larval stage must first acquire the virus and then pass it on to the grouse in the subsequent nymphal stage. While grouse and sheep are the only viraemic hosts, others can play an important role in the tick lifecycle and amplify the virus (Hudson et al. 1995). One of these is the mountain hare, *Lepus timidus*, which not only provides a suitable host for the female tick, but also allows nonviraemic transmission of the virus between cofeeding ticks (Jones et al. 1997). Thus, the level of exposure of grouse to the louping ill virus depends on the density of mountain hares in the region. Theoretically, it is possible for the virus to persist in the hare population alone in the total absence of grouse (Norman et al. 1999).

Louping ill mortality in red grouse occurs soon after the emergence of ticks in late spring and coincides with the production of young grouse, leading to significant mortality in young birds during early June. This, in turn, results in infected grouse having lower breeding success, lower bag records, and lower breeding densities of grouse (figure 6.7). Reduction in the production of young leads to a population growth rate that can be negative and, in some populations, numbers are only maintained by the immigration of grouse in the following winter (Hudson 1992). Not surprisingly, this heavy mortality and resultant low population density is followed by lower levels of nematode infection and, consequently, by a reduced tendency to cycle. In passing it should be noted that not all populations of grouse are exposed to ticks and not all tick populations carry the louping ill virus.

6.8 Predators

Grouse moor keepers are employed by the majority of privately owned estates to remove predators that take red grouse, principally the fox, *Vulpes vulpes*, and the carrion crow, *Corvus corone* (Hudson and Newborn 1995). Historically, this has included all predator species, although birds of prey and some of the threatened mammalian predators are now legally protected. Nevertheless, most keepers believe that birds of prey have a significant impact on grouse populations, so illegal killing of species like hen harriers, *Circus cyaneus*, still occurs.

An intensive and extensive investigation has recently been completed on the relationship between grouse and hen harriers. While there is no clear numerical response of harriers to grouse populations, they do exhibit a type III functional response when feeding on grouse and, as is well known, this could have a stabilizing effect on grouse population dynamics (Redpath and Thirgood 1999). The settlement of breeding harriers in relation to prey availability is positively related to the density of meadow pipits, indicating they breed in areas where pipits are abundant. Meadow pipits are a major food for the smaller males but, once established, the larger females tend to feed on red

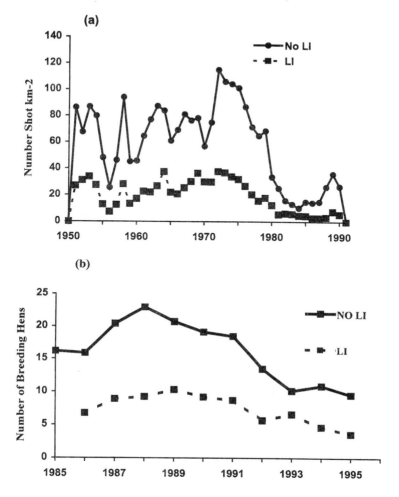

Figure 6.7 Population changes in red grouse infected with louping ill compared with uninfected populations. (a) Numbers harvested per annum per square kilometer. (b) Numbers of breeding adults.

grouse chicks (Redpath and Thirgood 1999). Following total protection of hen harriers on one estate, the population grew to 22 breeding harriers and the red grouse population fell. Although the initial decline was due to the regular grouse cycle, areas without harriers went through the normal increase phase while those with harriers continued to decline (figure 6.8). This is what is expected from a generalist predator with a type III functional response when the prey species becomes sparse—the "predator trap" or "predator pit."

Hudson et al. (1992a,b) examined the interaction between parasites, grouse, predators, and game keepers, and found an inverse relationship between keeper density and fox density, presumably because keepers killed

124 Population Cycles

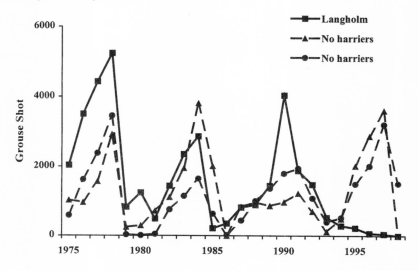

Figure 6.8 Numbers of grouse shot from the Langholm population following the protection of hen harriers in 1992 compared with two neighboring populations not exposed to harrier predation. The populations fluctuated in synchrony before the hen harriers became established, but subsequently the grouse numbers failed to rise at Langholm (after Thirgood et al. 1999).

most of them. In addition, the greater the density of keepers, the larger the harvest of grouse and the higher the intensity of worm infections. This implies that the removal of predators increases grouse populations and also results in higher parasite intensities.

To investigate whether parasites made the grouse more vulnerable to predation, dogs were used as artificial predators. Normally, trained pointing dogs can locate grouse at distances of up to and even more than 50 m, since grouse like many gamebirds produce a characteristic strong scent from their cecae. However, when female grouse start incubating young, they stop producing cecal feces, and dogs can walk within centimeters of them without detecting their presence. The cecae are also the site of infection by $T.\ tenuis$. Hudson et al. (1992b) postulated that parasitism may affect scent emission, making grouse vulnerable to predators during incubation. To test this hypothesis, trained pointing dogs were used to locate two groups of grouse. In one group, individuals had been randomly treated with an anthelmintic prior to the breeding season. In the other group, grouse were given water instead of the anthelmintic. Observers searching for nests by sight found them at random and in equal proportions, but dogs selectively found the nests of untreated red grouse, supporting the hypothesis that parasitic infections make the grouse more vulnerable to predators.

Incorporating selective predation into the model results in the removal of heavily infected grouse from the population, reduces the intensity of parasitism, and leads to an increase in the grouse population (figure 6.9a). In addition,

Parasitic Worms and Population Cycles of Red Grouse 125

increased selective predation dampens the cycles (figure 6.9b). Notice that the effect of predators is nonlinear, with low to moderate numbers having a positive effect on grouse equilibrium densities, but high numbers having a negative effect.

6.9 Harvesting

Hunting is another form of mortality that reduces the growth rate of grouse populations. Like predators and virus infections, we might expect harvesting

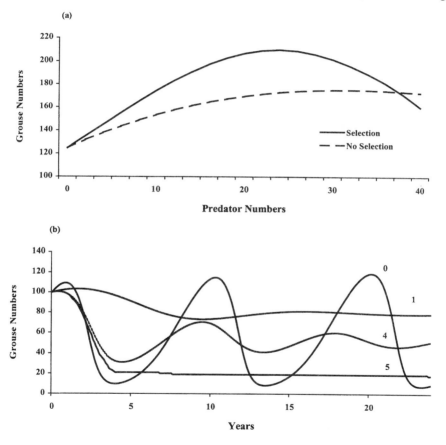

Figure 6.9 The influence of predation on the dynamics of red grouse populations as predicted by the grouse model. (a) Low to moderate rates of predation have a large impact on the worm population relative to their impact on the grouse population, leading to a release from worm-induced mortality and an increased grouse population. This is more marked when the predators selectively take the heavily infected grouse. (b) Simulations with the model illustrating the dampening effects of predation levels from 0 to 5.

to also have a stabilizing effect on grouse dynamics. The fact is, however, that grouse population fluctuations occur in the presence of hunting. The probable reason for this is that worm recruitment generally occurs just prior to the shooting season, so that hunting has little effect on the density-induced negative feedback responsible for the grouse cycle (Hudson 1986b).

6.10 Discussion and Summary

Populations of harvested red grouse exhibit cyclic fluctuations in abundance that vary in period and amplitude. We examined the hypothesis that the parasitic nematode *T. tenuis* increases numerically in response to grouse density, reduces the breeding success of female grouse and, through this delayed negative feedback loop, generates the cycles of abundance in grouse populations. Experimental studies of individual grouse confirmed that parasites reduce brood production. Realistic models of the parasite–host interaction generate cycles of similar period and amplitude to those observed in the field. Large-scale experiments conducted at the population level show that suppression of parasite infections greatly reduces the amplitude of the population cycle. Noncyclic populations were found to have lower infection levels, probably because habitat conditions reduced infection rates. Moderate levels of selective predation on heavily infected grouse dampens the cycles and increases the overall abundance of grouse. Natural enemies such as hen harriers and the tick-borne virus that causes louping ill disease reduce the survival of young grouse and tend to dampen oscillations. Harvesting does not seem to have this effect, probably because the period of worm recruitment occurs after chick mortality and before harvesting. None of the evidence discussed above is inconsistent with the hypothesis that population cycles in the red grouse are caused by delayed negative feedback between grouse and nematode parasite populations.

The significant role of parasites in causing red grouse cycles may well be a direct consequence of the activity of keepers protecting grouse from predators; that is, once predators are removed, the increasing population is subject to increasing levels of worm infection, which then cause reduced breeding production and a cyclic decline. Some workers consider this to be a rather exceptional case. However, there are a number of other cases where parasites are thought to play some role in population cycles. For example, there is increasing evidence that food quality interacting with parasitic nematodes is responsible for Soay sheep oscillations on St. Kilda (Gulland 1992); rock partridges exhibit unstable dynamics in northern Italy (Cattadori et al. 1999) that appear to be associated with infection by parasitic geohelminths (Rizzoli et al. 2000); population cycles of willow ptarmigan in Norway seem to be associated with the community of parasites that infect them (Holmstad and Skorping 1998); and, finally, unstable dynamics in Svalbard reindeer appear to be caused by the effects of gut helminths (Halvorsen and Bye 1999). The general pattern that emerges from all these examples is that there is a single host–parasite interaction where the parasite induces reduced breeding production. However, where other hosts play a significant role in the parasite lifecycle, the dynamics become more stable (Hudson and Greenman 1998).

This is not to say that red grouse cycles, or indeed cycles in any grouse species, are always caused by parasites. Such an extrapolation is risky at best. Unfortunately, there is currently no sound experimental evidence from other areas to test the hypotheses. There is, however, evidence that predators may play a significant role in some cyclic populations. For example, detailed studies over 16 years strongly suggest that the dynamics of rock ptarmigan in Iceland are destabilized by gyrfalcon predation (Nielsen 1999). What we can probably say with confidence is that one type of natural enemy, the parasites, can have an important destabilizing effect on some grouse populations, some of the time. Other natural enemies and resources, together with demographic features like spatial social structuring and age structure, can interact with stochastic and seasonal variations to influence the final pattern of fluctuation.

ACKNOWLEDGMENTS

We would like to thank all our scientific colleagues for their support and stimulating discussion on the subject of red grouse population cycles, especially Steve Redpath, Simon Thirgood, Bryan Grenfell, and Francois Mougeot. We also extend our gratitude to the landowners and keepers who allowed us to work on their land, and our warmest admiration to the dogs that actually did the fieldwork with enthusiasm and dedication. Finally we would also like to thank Robert Moss and Alan Berryman for their constructive comments and editorial suggestions.

REFERENCES

Anderson, R. M. and R. M. May. 1978. Regulation and stability of host–parasite interactions. I. Regulatory processes. *J. Anim. Ecol.* 47: 219–249.

Cattadori, I. M. C., P. J. Hudson, S. Merler, and A. P. Rizzoli. 1999. Temporal and spatial dynamics of rock partridge populations (*Alectoris graecae*) in northern Italy. *J. Anim. Ecol.* 68: 540–549.

Cobbold, T. S. 1873. *The grouse disease. The Field*, London. p. 15.

Dobson, A. P. and P. J. Hudson. 1992. Regulation and stability of a free-living host–parasite system, *Trichostrongylus tenuis* in red grouse. II. Population models. *J. Anim. Ecol.* 61: 487–498.

Dobson, A. P. and P. J. Hudson. The interaction between the parasites and predators of red grouse *Lagopus lagopus scoticus*. *Ibis* 137 (Suppl. 1): 87–96.

Gulland, F. M. D. 1992. The role of parasites in Soay sheep (*Ovis aries*) mortality during a population crash. *Parasitol.* 105: 493–503.

Gurney, W. S. C. and R. M. Nisbett. 1998. *Ecological dynamics*. Oxford University Press, Oxford.

Halvorsen, O. and K. Bye. 1999. Parasites, biodiversity and population dynamics in an ecosytem in the High Arctic. *Vet. Parasit.* 84: 205–227.

Holmstad, P.R. and A. Skorping. 1998. Covariation of parasite intensities in willow ptarmigan, *Lagopus lagopus* L. *Can. J. Zool.* 76: 1581–1588.

Hudson, P. J. 1986a. The effect of a parasitic nematode on the breeding production of red grouse. *J. Anim. Ecol.* 55: 85–94.

Hudson, P. J. 1986b. *The red grouse, the biology and management of a wild gamebird.* The Game Conservancy Trust, Fordingbridge, UK.

Hudson, P. J. 1992. *Grouse in space and time.* The Game Conservancy Trust, Fordingbridge, UK.

Hudson, P. J. and A. P. Dobson. 1996. Transmission dynamics and host–parasite interactions of *Trichostrongylus tenuis* in red grouse. *J. Parasitol.* 83: 194–202.

Hudson, P. J. and J. V. Greenman. 1998. Parasite mediated competition. Biological and theoretical progress. *Trends Ecol. Evol.* 13: 387–390.

Hudson, P. J. and D. Newborn. 1995. *A handbook of grouse and moorland management.* The Game Conservancy Trust, Fordingbridge, UK.

Hudson, P. J., A. P. Dobson, and D. Newborn. 1985. Cyclic and non-cyclic populations of red grouse: a role for parasitism? In D. Rollinson and R. M. Anderson (Eds.) *Ecology and genetics of host–parasite interactions.* Academic Press, London, pp 79–89.

Hudson, P. J., A. P. Dobson, and D. Newborn. 1992a. Do parasites make prey vulnerable to predation? Red grouse and parasites. *J. Anim. Ecol.* 61: 681–692.

Hudson, P. J., D. Newborn, and A. P. Dobson. 1992b. Regulation and stability of a free-living host–parasite system, *Trichostrongylus tenuis* in red grouse. I. Monitoring and parasite reduction experiments. *J. Anim. Ecol.* 61: 477–486.

Hudson, P. J., R. Norman, M. K. Laurenson, D. Newborn, M. Gaunt, H. Reid, E. Gould, R. Bowers, and A. P. Dobson. 1995. Persistence and transmission of tick-borne viruses: *Ixodes ricinus* and louping-ill virus in red grouse populations. *Parasitol.* 111: S49–S58.

Hudson, P. J., E. A. Gould, M. K. Laurenson, M. Gaunt, H. W. Reid, J. D. Jones, R. Norman, K. MacGuire, and D. Newborn. 1997. The epidemiology of louping-ill, a tick borne viral infection of grouse and sheep. *Parasitologia* 39: 319–323.

Hudson, P. J., A. P. Dobson, and D. Newborn. 1998. Prevention of population cycles by parasite removal. *Science* 282: 2256–2258.

Hudson, P. J., A. P. Dobson, and D. Newborn. 1999. Population cycles and parasitism. *Science* 286: 2425.

Jenkins, D., A. Watson, and R. G. Miller. 1963. Population studies on red grouse, *Lagopus lagopus scoticus* (Lath.) in north-east Scotland. *J. Anim. Ecol.* 32: 317–376.

Jones, L. D., M. Gaunt, R. S. Hails, K. Laurenson, P. J. Hudson, H. Reid, P. Henbest, and E. A. Gould. 1997. Efficient transfer of louping ill virus between infected and uninfected ticks cofeeding on mountain hares (*Lepus timidus*). *Med. Vet. Entomol.* 11: 172–176.

Lack, D. 1954. *The natural regulation of animal numbers.* Oxford University Press, Oxford.

Lambin, X., C. J. Krebs, R. Moss, N. C. Stenseth, and N. G. Yoccoz. 1999. Population cycles and parasitism. *Science* 286: 2425.

Laurenson, M. K., P. J. Hudson, K. McGuire. S. J. Thirgood, and H. W. Reid. 1998. Efficacy of acaricidal tags and pour-on as prophylaxis against ticks and louping-ill in red grouse. *Med. Vet. Entomol.* 11: 389–393.

Lovat, L. 1911. *The grouse in health and disease.* Smith, Elder and Co., London, p. 512.

MacDonald, D. G. F. 1883. *Grouse disease: its causes and remedies.* W. H. Allen and Co., London.
Mackenzie, J. M. D. 1952. Fluctuations in the numbers of Tetraonidae. *J. Anim. Ecol.* 21: 128–153.
May, R. M. 1976. *Theoretical ecology. Principles and applications.* Blackwell Scientific, Oxford.
May, R. M. and R. M. Anderson. 1978. Regulation and stability of host–parasite interactions. II. Destabilizing. *J. Anim. Ecol.* 47: 249–268.
Moran, P. A. P. 1952. The statistical analysis of gamebird records. *J. Anim. Ecol.* 21: 154–158
Moss, R. and A. Watson. 1991. Population cycles and kin selection in red grouse *Lagopus lagopus scoticus. Ibis* Suppl. 1: 113–120.
Nielsen, O. K. 1999. Gyrfalcon predation on ptarmigan: numerical and functional responses. *J. Anim. Ecol.* 68: 1034–1050.
Norman, R., R. G. Bowers, M. E. Begon, and P. J. Hudson. 1999. Persistence of tick borne virus in the presence of multiple host species: tick reservoirs and parasite mediated competition. *J. Theor. Popul. Biol.* 200: 111–118.
Potts, G. R., S. C. Tapper, and P. J. Hudson. 1984. Population fluctuations in red grouse: analysis of bag records and a simulation model. *J. Anim. Ecol.* 53: 21–36
Redpath, S. M. and S. J. Thirgood. 1999. Numerical and functional responses of generalist predators: harriers and *peregerines* on grouse moors. *J. Anim. Ecol.* 68: 879–892.
Reid, H. W., J. S. Duncan, J. D. P. Phillips, R. Moss, and A. Watson. 1978. Studies of louping ill virus in wild red grouse (*Lagopus lagopus scoticus*). *J. Hygiene* 81: 321–329
Ricklefs, R. E. 1979. *Ecology.* Thomas Nelson, Melbourne.
Rizzoli, A., P. J. Hudson, M. T. Manfredi, F. Rosso, and I. M. C. Cattadori. 2000. Intensity of nematode infections in cyclic and non-cyclic rock partridge populations. *Parasitologia* 41: 561–565.
Saunders, L. M., D. Tompkins, and P. J. Hudson. 1999. Investigating the dynamics of nematode transmission to the red grouse (*Lagopus lagopus scoticus*): studies on the recovery of *Trichostrongylus tenuis* larvae from vegetation. *J. Helminthol.* 73: 171–175.
Saunders, L. M., D. Tompkins, and P. J. Hudson. 2000. Spatial aggregation and temporal migration of free-living stages of the parasitic nematode *Trichostrongylus tenuis. Funct. Ecol.* 14: 468–473.
Shaw, J. L. 1988. Arrested development in *Trichostrongylus tenuis* as third stage larvae in red grouse. *Res. Vet. Sci.* 48: 256–258.
Shaw, D. J. and A. P. Dobson. 1995. Patterns of macroparasite abundance and aggregation in wildlife populations: a quantitative review. *Parasitol.* 111: 111–134.
Thirgood, S. J., D. Haydon, P. Rothery, P. Redpath, I. Newton, and P. J. Hudson. 1999. Habitat loss and raptor predation: disentangling long- and short-term causes of red grouse declines. *Proc. Roy. Soc.* 267: 651–656.
Watson, H., D. L. Lee, and P. J. Hudson. 1987. The effect of *Trichostrongylus tenuis* on the caecal mucosa of young, old and anthelmintic treated wild red grouse *Lagopus lagopus scoticus. Parasitol.* 94: 405–411.
Wynne-Edwards, V. C. 1986. *Groups selection.* Blackwell Scientific, Oxford.

7

Population Cycles of the Larch Budmoth in Switzerland

Peter Turchin, Cheryl J. Briggs, Stephen P. Ellner,
Andreas Fischlin, Bruce E. Kendall, Edward McCauley,
William W. Murdoch, and Simon N. Wood

7.1 Introduction

The population dynamics of the larch budmoth (LBM), *Zeiraphera diniana*, in the Swiss Alps are perhaps the best example of periodic oscillations in ecology (figure 7.1). These oscillations are characterized by a remarkably regular periodicity, and by an enormous range of densities experienced during a typical cycle (about 100,000-fold difference between peak and trough numbers). Furthermore, nonlinear time series analysis of LBM data (e.g., Turchin 1990, Turchin and Taylor 1992) indicates that LBM oscillations are definitely generated by a *second-order* dynamical process (in other words, there is a strong delayed density dependence—see also chapter 1). Analysis of time series data on LBM dynamics from five valleys in the Alps suggests that around 90% of variance in R_t is explained by the phenomenological time series model employing lagged LBM densities, $R_t = f(N_{t-1}, N_{t-2})$ (Turchin 2002).

As discussed in the influential review by Baltensweiler and Fischlin (1988) about a decade ago, ecological theory suggests a number of candidate mechanisms that can produce the type of dynamics observed in the LBM (see also chapter 1). Baltensweiler and Fischlin concluded that changes in food quality induced by previous budmoth feeding was the most plausible explanation for the population cycles. During the last decade, the issue of larch budmoth oscillations was periodically revisited by various population ecologists looking for general insights about insect population cycles (e.g., Royama 1977, Bowers et al. 1993, Ginzburg and Taneyhill 1994, Den Boer and Reddingius 1996, Hunter and Dwyer 1998, Berryman 1999). These

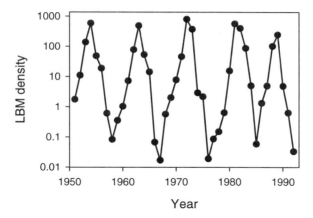

Figure 7.1 Population oscillations of the larch budmoth at Sils (Upper Engadine Valley, Switzerland). Moth density is the number of larvae per kilogram of larch branches (data from Baltensweiler 1993b).

authors generally concurred with the view that budmoth cycles are driven by the interaction with food quality. A recent reanalysis of the rich data set on budmoth population ecology collected by Swiss researchers over a period of several decades, however, suggested that the role of parasitism is underappreciated (Turchin et al. 2002).

7.2 General Overview of Hypotheses for LBM Cycles

Before focusing on the roles of food quality and parasitism in LBM dynamics, we briefly review the status of other hypotheses that were discussed in the literature on LBM cycles. First, the natural history of the LBM–larch system is such that *food quantity* is an unlikely factor to explain LBM oscillations. Mortality of the host trees due to defoliation is less than 1% (Baltensweiler and Fischlin 1988). Although the length of needles is reduced after a severe defoliation, the total amount of needle biomass is decreased only about two-fold. It is difficult to imagine how such small variations in food availability could drive a second-order population cycle in which the ratio of peak/trough densities is around 100,000. However, food quantity should act as a strong first-order mechanism regulating LBM density, since most LBM outbreaks are accompanied by widespread defoliation of host trees (Baltensweiler and Fischlin 1988), leading to mass starvation of larvae during peak years.

Maternal effects constitute a theoretically plausible intrinsic mechanism of second-order oscillations (Ginzburg and Taneyhill 1994). However, there is no evidence that this mechanism operates in the LBM. Even the proponents of the maternal effect hypothesis admit that their model is not applicable to the LBM (Ginzburg and Taneyhill 1994).

Theory suggests that the interaction between *pathogens* and their hosts can exhibit oscillatory dynamics (e.g., Anderson and May 1980). In 1957, after the first cycle that was studied intensively, it seemed obvious to everybody that a granulosis virus disease played a critical role in suppressing the outbreak (Baltensweiler and Fischlin 1988). Unfortunately, the incidence of virus decreased during the next outbreak, and then disappeared completely. As a result, the pathogen hypothesis fell out of vogue, at least among field workers. Despite this, Anderson and May (1980) used LBM as their prime example of how an epidemiological model may explain population cycles in a forest insect.

There are two larch budmoth host races with distinct differences in color and ecological traits—a dark morph that feeds primarily on deciduous larch, and a light morph that feeds primarily on evergreens (*Pinus cembra* and *Picea abies*). The frequency of the dark morph tends to increase during outbreaks and decrease during declines (Baltensweiler 1993a, fig. 1). Baltensweiler (1977, 1993a) proposed the following explanation for this pattern: During population increases, the dark morphs become more abundant because they have faster development and higher survival than the light morphs. During population collapses, the dark morphs decrease faster than the light morphs, because they rely primarily on larch for food, and the quality of larch foliage is reduced by defoliation. Once the effects of defoliation on host quality dissipate, dark morphs begin increasing faster than light ones, and the cycle repeats itself.

Baltensweiler (1993a) argued that this polymorphism plays a key role in the LBM cycle. In particular, he suggested that it helps explain why low LBM populations switch immediately from the decline to the increase phase. However, as we shall see later, the abrupt switch from decline to increase is not a pattern that needs a special explanation because it arises naturally in several models considered later. Furthermore, the polymorphic fitness hypothesis is not an elemental mechanism, because it invokes plant quality as the primary factor causing population collapse (without prolonged decrease in plant quality the population density of dark morphs would not decrease, and no cycle would ensue). Thus, the polymorphic fitness hypothesis is not an explanation of the primary question (why LBM populations oscillate), but rather why morph frequencies change regularly during the LBM cycle. It is a consequence not a cause of the cycle.

The *food quality* hypothesis is currently the dominant explanation of LBM oscillations (Baltensweiler and Fischlin 1988). Larch trees suffering greater than 50% defoliation lack nutrient resources to grow high-quality needles during the following spring. Needles grown after the LBM peak are short (< 20 mm, compared with a normal length of > 30 mm) and have a high raw fiber content of about 18% (compared with the normal 12%), while the raw protein content falls from 6% to 4%. Low quality of food (as measured by high raw fiber, and indexed by low needle length) strongly depresses larval survival and female fecundity (Benz 1974, Omlin 1977). Furthermore, poor needle quality persists for several years after an outbreak. This "quality transmission" effect imposes delayed density dependence on LBM population growth rates, and can theoretically lead to cycles, as shown by the model developed by Fischlin (1982; see also Fischlin and Baltensweiler 1979).

General theory suggests that *parasitoids* may play an important role in population dynamics of forest insects, and LBM parasitoids were intensively studied from the beginning of the systematic research program (e.g., Baltensweiler 1958). Once the data on parasitism rates became available, however, the initial enthusiasm for the parasitoid hypothesis waned. Parasitism rates at the population peak are typically low, around 10–20% (Baltensweiler and Fischlin 1988), suggesting that parasitoids play a minor role in *limiting* LBM densities; that is, in preventing further LBM increases. The parasitism rate reaches a high of around 90% during the collapse stage, but this high is reached only during the second (or even third) year after the peak. Accordingly, Delucchi (1982) concluded that control of LBM by parasitoids alone is not possible, and the current thinking is that parasites merely track the larch budmoth population; that is, budmoth fluctuations regulate the number of parasitoids and not vice versa. However, the observation that parasitoids do not play an important role in stopping LBM increases does not necessarily mean that they are a minor agent in LBM dynamics. This conclusion is erroneous because it confuses first- and second-order factors; that is, a mechanism imposing an upper bound on LBM population density may differ from one inducing oscillations.

In summary, there are two hypotheses about LBM cycles that require further examination—the plant quality hypothesis and the parasitism hypothesis. In the following sections we review the data, the models, and especially the cross-connections between empirical and theoretical approaches relevant to each of the hypotheses.

7.3 LBM–Plant Quality Interaction

Previous analyses of the interaction between plant quality and LBM dynamics emphasized the raw fiber content of larch needles as the main indicator of food quality (for example, the model of Fischlin was based on this index). However, there is no time series data available for this index, while we have an extensive data set for another index, the average needle length. Before using these data, however, we first need to check on how well needle length reflects the food quality from the point of view of LBM larvae. We can answer this question with the bioassay data of Benz (1974, table 8). Benz fed LBM larvae foliage from larch trees with known defoliation history, and measured larval survival and pupal weight. Because female pupal weight is linearly related to fecundity, we can translate the measured pupal weights into expected fecundity using the relationship estimated by Benz (1974, fig. 2).

Multiplying larval survival by the calculated fecundity we then obtain a measure related to the finite rate of population increase λ' (the prime is to remind us that this measure is not the true λ because it does not include egg and adult mortality). Plotting λ' against needle length index reveals a well-defined relationship between these two quantities, with a high coefficient of determination, $r^2 = .86$ (fig. 7.2). Interestingly, the alternative index, raw fiber content, explains a somewhat lower percentage of variance in λ' ($r^2 = .66$; analysis based on the same Benz data). Thus, the somewhat surprising conclusion is that needle length appears to be a better index of food quality than raw fiber content. Clearly, food quality is a complex variable, whose effect on LBM survival and fecundity is mediated by physical (e.g., toughness as measured by raw fiber content) and nutritional (e.g., protein content) properties of needles, as well as, perhaps, tree chemical defenses; for example, resin content (Benz 1974). However, the observation that the average length of needles is an accurate predictor of LBM rates of population change is encouraging.

We now consider the results of the analysis of time series data on LBM density and needle length during 1961–92 at Sils (Engadine Valley, Switzerland) (see figure 7.3a). Turchin et al. (2002) employed nonlinear regression to investigate the cross-effects of LBM density and needle length on each other. Although we tried a variety of functional forms for the general model $R_t = f(N_{t-1}, Q_{t-1})$ (where $R_t = \ln N_t/N_{t-1}$ is the realized per-capita rate of the budmoth population, and N_t and Q_t are LBM density and needle length in year t), we could detect no strong effect of needle length (less than one third of the variance explained).

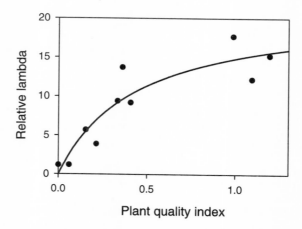

Figure 7.2 Effect of plant quality on the relative multiplication rate of the larch budmoth, k (calculations based on data from Benz 1974, table 8). Plant quality index is scaled by needle length: $Q_t =$ (needle length $-$ 15 mm)/15 mm (this scales the index to the range of approximately 0–1).

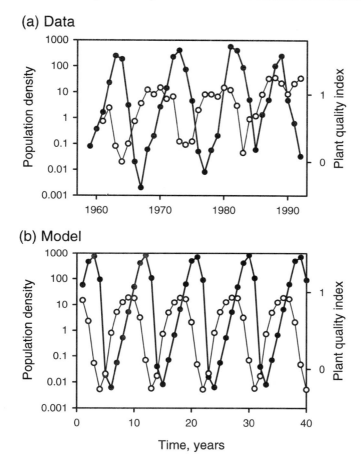

Figure 7.3 Dynamics of LBM density and food quality: (a) observed and (b) predicted by the model.

By contrast, the dynamics of needle length are strongly related to LBM density and the previous year's index. The following simple model (which is a discretization of Edelstein-Keshet and Rausher 1989; see Turchin 2002):

$$Q_t = (1-\alpha) + \alpha Q_{t-1} + \frac{cN_{t-1}}{d + N_{t-1}} \tag{7.1}$$

explained around 75% of the variance in Q_t. The effect of "memory," represented by the autoregressive parameter α, is highly significant, and by itself explains about 47% of the variance.

The surprising result from these analyses is that an index of plant quality explains a very low proportion of variance in the LBM rate of change. Such a low level of predictability associated with plant quality contrasts unfavorably with the r^2 of around 90% suggested by phenomenological response-surface models, or regression analyses utilizing parasitism data (see section 7.4). While these regression analyses do not consitute a "proof" that plant quality is unimportant in LBM dynamics, they considerably weaken the case for it as the primary factor responsible for LBM oscillations. However, in order to pass the final verdict, we need to know whether a model based on the plant quality hypothesis is capable of mimicking the observed LBM dynamics.

In constructing the equation for LBM dynamics, we used the data depicted in figure 7.2. After trying several two-parameter relationships, we found that a negative exponential function fits the data best (this is a purely phenomenological approach, as we have no mechanistic basis for postulating a functional form). Using log-transformed λ' ($R' = \log \lambda'$) as the response variable, the fitted curve was

$$R'_t = a[1 - \exp(Q_{t-1}\delta)], \tag{7.2}$$

where $a = 3.8$ is the maximum rate of increase at the highest food quality, and $\delta = 0.22$ is the parameter determining how fast the rate of change approaches its maximum. There are two things still missing from this model. First, it assumes that there is no mortality in the adult and small larva stages. We can remedy this by replacing the maximum rate a with the average per-capita rate of population change observed when plant quality is at its highest. A good choice for this parameter is $R_0 = 2.5$, corresponding to about a 10-fold increase in N_t per year (because this is the average rate at which the LBM density climbs out of the trough). Second, the model lacks a self-limitation term due to larvae overeating their food supply and starving as a result. One solution is to use the Ricker equation, which leads to the following model for LBM dynamics:

$$N_{t+1} = N_t \exp\{R_0[1 - \exp(Q_t/\delta)] - R_0 N_t/K\}. \tag{7.3}$$

For the dynamics of needle length, Q_t, we simply use the empirical equation (7.1). The regression-based parameter estimates (mean ± SE) are $\delta = 0.22 \pm 0.05$, $\alpha = 0.5 \pm 0.1$, $c = 0.7 \pm 0.2$, and $d = 150 \pm 150$. Additionally, we have $R_0 = 2.5 \pm 0.2$ and $K = 500 \pm 200$.

Numerical exploration of dynamics for parameters in the ranges defined by mean ± SE indicated that this model is readily capable of generating population trajectories resembling the data (figure 7.3b). Trajectories predicted by the model match both the period and the amplitude of the observed LBM oscillations. Additionally, the model mimics the quantitative pattern of the quality index dynamics reasonably well, including the amplitude of variation and the timing of declines and increases (compare with figure 7.3a). However, the range of oscillations in Q_t predicted by the model is somewhat lower than that observed.

In summary, the model of LBM–plant quality interactions, with biologically plausible parameters, is capable of matching the empirically observed quantitative patterns in the time series data. Does it mean that we have found the explanation for the LBM oscillations? Unfortunately, there remains one serious problem, the lack of detectable effect of Q_t on the LBM rate of change; that is, no negative feedback between Q_t and N_t.

Furthermore, if we examine the last documented LBM outbreak (peak in 1989), we notice that the plant quality index hardly declined at all, with needle lengths remaining at high levels through the whole period (figure 7.3a). As discussed by Baltensweiler (1993b), a sequence of unusual weather in 1989–91 caused high egg mortality. As a result, the budmoth population never reached the level at which widespread defoliation occurs (the 1989 peak density was only 240 larvae per kilogram of larch branches, while previous peak densities observed at Sils were 490, 590, 800, and 560 larvae/kg). Correspondingly, light defoliation resulted in no decline in plant quality. Yet, the LBM population collapsed during 1990–92. In other words, we have here a natural experiment suggesting that a large decrease in plant quality is not necessary for LBM cycles.

7.4 Parasitism Hypothesis

Our investigation of the parasitism hypothesis employs an approach similar to that used in assessing the plant quality hypothesis. First, we subject time series data to nonlinear regression analyses. Then, we develop an empirically based model of LBM–parasitoid interaction that attempts to mimic the observed dynamics.

The general model that we used (Turchin et al. 2002) was based on the Nicholson–Bailey framework, to which we added a self-limitation term in the host and a Beddington-type functional response (this general form of functional response incorporates both the effects of handling time h and mutual interference between parasitoids, parameterized by wasted time w). The equations were:

$$N_{t+1} = N_t \exp[R_0(1 - N_t/K) - aP_t/(1 + ahN_t + awP_t)], \qquad (7.4a)$$

$$P_{t+1} = N_t S_t, \quad \text{where} \quad S_t = 1 - \exp[-aP_t/(1 + ahN_t + awP_t)]. \qquad (7.4b)$$

The parasitoid density, P_t, is not directly observed, and therefore, we need to estimate it by multiplying the host density during the previous year by that year's parasitism rate: $P_t = N_{t-1}S_{t-1}$ (S_t is the proportion of hosts parasitized in year t). Note that our estimate of P_t does not incorporate the (unknown) overwintering mortality. Thus, P_t is actually a relative index that is linearly related to the true parasitoid density, but with an unknown proportionality constant (this has no effect on the estimate of the proportion of variance resolved by parasitism).

Results of nonlinear regression suggest that the parasitism rate is quite well resolved by model (7.4). Thus, the simple three-parameter equation (7.4b) resolves 71% of the variance in the parasitism rate. The coefficient of determination for the LBM rate of change is even higher, with equation (7.4a) resolving 88% of the variance. What is particularly impressive is that a very simple three parameter model:

$$R_t = \ln(N_t/N_{t-1}) = R_0 - aP_{t-1}/(1 + awP_{t-1}) \tag{7.5}$$

manages to capture almost as high a proportion of variance, $r^2 = .86$.

To summarize, a simple, but theoretically sound, model based on the parasitism hypothesis resolves close to 90% of the variation in the LBM rate of change. The regression analysis suggests that model (7.4) can be simplified by setting parameter h to 0, because this procedure does not decrease the percentage of variance explained by the parasitism model.

The regression analysis also yields estimates of parameters $a = 2.5 \pm 1$ and $w = 0.17 \pm 0.02$ [means $\pm$ SE estimated by fitting equation (7.5) to the data]. We have already estimated R_0 and K above ($R_0 = 2.5 \pm 0.2$ and $K = 250 \pm 50$). Simulating the model within these parameter ranges shows that it produces high-amplitude oscillations for all reasonable values of parameters. For the median parameter values, however, the period is a bit short—7 years compared with the observed 8–9-year period. It is necessary to reduce w to 0.15 in order to lengthen the period to 8 years, and to 0.13 (2 SE from the point estimate and still within the realm of the possible) in order to lengthen the period further to 9 years. The model output matches well other characteristics of the data of that period, such as the amplitude and the cross-correlation function between LBM density and the proportion parasitized. In particular, the proportion parasitized peaks on average 2 years after the LBM peak, similarly to the pattern observed in the data.

7.5 Putting It All Together: A Parasitism-Quality Model

The preceding analyses of data and models suggest an interesting conclusion. On the one hand, the model with plant quality as the only mechanism driving second-order oscillation fails to match data patterns as well as the LBM–parasitoid model. On the other hand, short-term experiments suggest that there is a strong effect of changes in plant quality on LBM survival and reproduction. This raises an important question: Should we be satisfied with the parasitism-only explanation of the LBM dynamics, or do we instead need a multifactorial model, combining plant quality and parasitism? One way to address this issue is to investigate the dynamics predicted by the multifactorial model, and contrast its ability to match empirical patterns with the two simpler alternatives.

Combining the effects of plant quality and parasitism is quite straightforward, now that we have invested so much effort in building models for each component separately. The equations of this parasitism–quality model are:

$$Q_{t+1} = (1-\alpha) + aQ_t - cN_t/(d+N_t), \quad (7.6a)$$

$$N_{t+1} = N_t \exp\{R_0[1-\exp(Q_t/\delta)] - R_0 N_t/K - aP_t/(1+awP_t)\} \quad (7.6b)$$

$$P_{t+1} = N_t S_t, \quad \text{where} \quad S_t = 1 - \exp[-aP_t/(1+awP_t)]. \quad (7.6c)$$

Parameter estimates are the same as above. Simulating the model within these parameter ranges, we find that the model does very well for parameters at their median values (or very near to them). In particular, with slight modifications (specifically, $R_0 = 2.3$, $c = 0.9$, and $d = 100$; note that with each of these modifications we are staying within 1 SE of the median estimates), the model output matches the data patterns very well (figure 7.4). Quantitative measures of the observed time series pattern (periodicity, amplitude, and cross-correlations between LBM and parasitism or quality index) are also closely matched by the model-generated trajectories.

7.6 Discussion

Our theoretical and empirical analyses suggest the following conclusions. First, our reanalysis supports the previous efforts modeling the LBM–plant food quality interaction. A simple model with biologically plausible parameters (in fact, estimates based on experimental data) predicts population dynamics that are quite similar to the observed pattern (including matching such quantitative characteristics of observed fluctuations as order, periodicity, and amplitude). However, the plant quality hypothesis has weaknesses: Although the model predicts that there should be a strong feedback effect

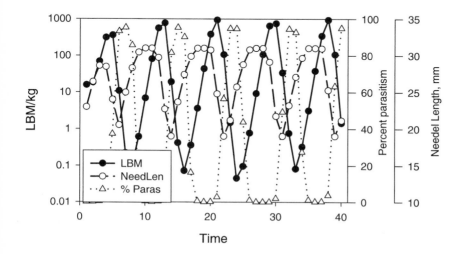

Figure 7.4 Dynamics of LBM density, food quality, and parasitism rate predicted by the tritrophic model.

from plant quality to the LBM rate of change, analysis of real data does not reveal it. Additionally, lack of quality decline during the last recorded cyclic collapse is hard to understand if plant quality is the main factor driving LBM oscillations.

Second, a model based on the parasitism hypothesis with empirical estimates of parameters is capable of mimicking the observed LBM dynamics. In this it is similar to the plant quality hypothesis. Unlike the rival explanation, however, the parasitism hypothesis is supported by regression analyses of the feedback relationship from parasitism rates to the LBM rate of change. However, to match the observed period, we have to "stretch" some parameter values. Additionally, the experimentally observed effect of plant quality is rather strong (at least 10-fold change in λ as a function of quality). It is generally a bad modeling approach to add a mechanism to the model simply because "it is there." Having such a strong numerical effect, however, makes one wonder whether the explanation of LBM cycles should leave it out.

Finally, a tritrophic model combining both hypotheses does the best job at matching the observed dynamics for biologically reasonable parameter values. We have, thus, an interesting situation. On epistemological grounds, the tritrophic hypothesis suffers because it is more complex than either of the plant quality or parasitism hypotheses. However, while both simple hypotheses can reproduce the fluctuation pattern of the primary data (LBM density), they fail in some other ways, as discussed above. Therefore, given the present state of knowledge, we conclude that the currently best-supported hypothesis is that LBM cycles are driven by *both* parasitism and plant quality interactions.

If this conclusion is correct, then parasitoids and plant quality act as a *dynamical complex*. This means that if, for whatever reason, one factor fails to cause a prolonged LBM density collapse after peak, then the other factor ensures that the cycle will continue, as apparently happened in the early 1990s. Assuming that this example of multiple causation is not unique to larch budmoth, it raises an important methodological issue. It is possible that by experimentally manipulating each factor we may "prove" that it is not the "necessary condition" for sustained cycles, leading to an erroneous rejection of both hypotheses. Only an experiment that manipulates both factors together (ideally coupled with a modeling program) can, in principle, resolve the question of what mechanisms drive population oscillations (see, e.g., chapter 4).

REFERENCES

Anderson, R. M. and R. M. May. 1980. Infectious diseases and population cycles of forest insects. *Science* 210: 658–661.

Baltensweiler, W. 1958. Zur Kenntnuis der Parasiten des Grauen Larchenwichlers (*Zeiraphera griseana* Hubner) in Oberengadin. *Mitt. Eidgen. Anst. forst. Versuch.* 34: 399–477.

Baltensweiler, W. 1977. Colour-polymorphism and dynamics of larch budmoth populations (*Zeiraphera diniana* Gn., Lep. Tortricidae). *Mitt. Schweiz. Entomol. Ges.* 50: 15–23.

Baltensweiler, W. 1993a. A contribution to the explanation of the larch bud moth cycle, the polymorphic fitness hypothesis. *Oecologia* 93: 251–255.

Baltensweiler, W. 1993b. Why the larch bud-moth cycle collapsed in the subalpine larch–cembran pine forests in the year 1990 for the first time since 1850. *Oecologia* 94: 62–66.

Baltensweiler, W. and A. Fischlin. 1988. The larch budmoth in the Alps. In A. A. Berryman (Ed.) *Dynamics of forest insect populations: patterns, causes, implications.* Plenum Press, New York, pp. 331–351.

Benz, G. 1974. Negative Ruckkopelung durch Raum- und Nahrungskonkurrenz sowie zyklische Veranderung der Nahrungsgrundlage als Regelsprinzip in der Populationsdynamik des Grauen Larchenwicklers, Zeiraphera diniana (Guenee) (Lep. Tortricidae). *Z. Angew. Entomol.* 76: 196–228.

Delucchi, V. 1982. Parasitoids and hyperparasitoids of *Zeiraphera diniana* (Lep. Tortricidae) and their role in population control in outbreak areas. *Entomophaga* 27: 77–92.

Edelstein-Keshet, L. and M. D. Rausher. 1989. The effects of inducible plant defenses on herbivore populations. I. Mobile herbivores in continuous time. *Am. Nat.* 133: 787–810.

Fischlin, A. 1982. *Analyse eines wald-insekten-systems: der subalpine Lärchenarvenwald und der graue Lärchenwickler* Zeiraphera diniana *Gn. (Lep. Tortricidae).* Ph.D. thesis no. 6977, ETH, Zurich.

Fischlin, A. and W. Baltensweiler. 1979. Systems analysis of the larch bud moth system. Part 1. the larch–larch bud moth relationship. *Mitt. Schweiz. Entomol. Ges.* 52: 273–289.

Ginzburg, L. R. and D. E. Taneyhill. 1994. Population cycles of forest Lepidoptera: a maternal effect hypothesis. *J. Anim. Ecol.* 63: 79–92.

Omlin, F. X. 1977. *Zur populationsdynamischen Wirkung der durch Raupenfrass und Dungung veranderten Nahrungsbasis auf den Grauen Larchenwickler* Zeiraphera diniana *Gn. (Lep. Tortricidae).* Ph.D. thesis no. 6064, ETH, Zurich.

Turchin, P. 1990. Rarity of density dependence or population regulation with lags? *Nature* 344: 660–663.

Turchin, P. 2002. *Complex population dynamics: a theoretical/empirical synthesis.* Princeton University Press, Princeton, N.J.

Turchin, P. and A. D. Taylor. 1992. Complex dynamics in ecological time series. *Ecology* 73: 289–305.

Turchin, P., S. P. Ellner, S. N. Wood, B. E. Kendall, W. W. Murdoch, A. Fischlin, J. Casas, E. McCauley, and C. J. Briggs. 2002. Dynamical effects of plant quality and parasitism on population cycles of larch budmoth. *Ecology* (accepted).

8

Population Cycles of the Autumnal Moth in Fennoscandia

Miia Tanhuanpää, Kai Ruohomäki, Peter Turchin, Matthew P. Ayres, Helena Bylund, Pekka Kaitaniemi, Toomas Tammaru, and Erkki Haukioja

8.1 Introduction

Most species of insect herbivores are restricted to low densities, but some display large-scale density fluctuations, including periodic outbreaks (Faeth 1987, Mason 1987, Hanski 1990, Hunter 1995). The tendency to reach high densities has been related to certain life history traits (Hunter 1991, 1995, Tammaru and Haukioja 1996). However, all populations of a given outbreaking species do not necessarily display high densities. In those cases, outbreaks are frequently more pronounced in populations in physically severe and marginal habitats (Wallner 1987, Myers and Rothman 1995).

The autumnal moth, *Epirrita autumnata* (Borkhausen) (Lepidoptera: Geometridae) is an example of a species with both outbreaking and nonoutbreaking populations. In mountain birch [*Betula pubescens* ssp. *czerepanovii* (Orlova) Hämet-Ahti] forests of northern and mountainous Fennoscandia (hereafter northern populations), *E. autumnata* displays fluctuations with a statistically significant periodicity of 9–10 years (Tenow 1972, Haukioja et al. 1988, Bylund 1995). During outbreaks, forests may be totally defoliated and trees may even die over large areas (Tenow 1972, Lehtonen and Heikkinen 1995). In more southern parts of the species' Holarctic distribution (hereafter southern populations), outbreaks are absent and populations remain at low densities.

Cycles of northern *E. autumnata* populations vary in their amplitude (Tenow 1972). Outbreak densities that produce conspicuous defoliation are typically reached in only some areas, and often in different areas during successive peaks (Tenow and Bylund 1989). Empirical data indicate a fairly regular pattern of fluctuations, that is synchronous on a regional scale, also in populations with moderate or low peak densities (Bylund 1997). Thus, there are two main questions regarding population regulation of northern and mountainous *E. autumnata*—what causes the cycles, and what causes spatial variations in outbreak severity? In southern populations, the main question is what prevents outbreaks?

8.2 The Species

Larvae of *E. autumnata* hatch early in spring at the time of birch bud break. Birches (*Betula* spp.) are the main host plants, although larvae are able to feed on many deciduous trees and shrubs (Seppänen 1970). The green cryptic larvae feed freely within the canopy of their host plant. In midsummer, after five larval instars, they descend to the ground and pupate. Short-lived nocturnal adults fly in autumn. Adult moths do not normally feed (Tammaru et al. 1996), so fecundity is directly dependent on body mass accumulated during the larval stage (Tammaru et al. 1996, Tammaru 1998). Females lay eggs mostly singly among lichens on tree branches where they overwinter (Tenow and Bylund 1989, Tammaru et al. 1995). According to Kaitaniemi et al. (1999), the average maximum fecundity is 130 eggs per female in northern populations, and this gives them a high potential growth rate ($r_m > 4$; Haukioja et al. 1988).

8.3 What Causes Population Cycles?

8.3.1 Physical Effects

In some systems, density-independent factors, such as periodic climate, may be largely responsible for causing the fluctuations of populations (e.g., Leirs et al. 1997, Speight et al. 1999, Turchin and Berryman 2000). The solar cycle is a factor resulting in periodic climate that has approximately the same cycle lengths as *E. autumnata* (see Waldmeier 1961). In *E. autumnata*, the increase phases of the cycles are associated with the decline phases of the solar cycle (Ruohomäki et al. 2000). A plausible causality between the solar cycle and *E. autumnata* population cycles is temperature, as it covaries with both sunspot activity and *E. autumnata* performance. However, as the causality between the solar cycle and population cycles is practically impossible to assess with present statistical tools (Lindström et al. 1996), the hypothesis that the solar cycle is the factor driving population cycles remains to be proven or falsified.

8.3.2 Maternal and Genetic Effects

There is no evidence for phase-dependent maternal or genetic differences in reproductive capacity, flight ability, or host plant use in the autumnal moth

(Ruohomäki 1992, Ruohomäki and Haukioja 1992a,b), and thus, intrinsic factors are apparently not strongly involved in the population cycles.

8.3.3 Trophic Interactions

For trophic interactions to produce cyclic density fluctuations, delayed negative feedback must be involved (e.g., chapter 1, Berryman 1981, Turchin 1990, Sinclair and Pech 1996). Two such factors in the autumnal moth system, identified from empirical studies, are larval parasitoids and delayed induced responses of mountain birch trees.

8.3.3.1 Larval Parasitism

Specialist parasitoids have been suggested to cause cyclic density fluctuations in populations of forest Lepidoptera (e.g., Berryman 1996). At least seven species of hymenopterous parasitoids attack northern *E. autumnata* larvae (Ruohomäki 1994). All species also parasitize other caterpillars, but alternative hosts are generally rare in the subarctic region. Hence, all parasitoids apparently function as specialists (*functional specialists*) on northern *E. autumnata*. Parasitism rates range from 0 (during the increase phase of the autumnal moth cycle) to close to 100% (during the decline) (Bylund 1995, Berryman 1996). Long-term monitoring of larval densities and parasitism rates (Ruohomäki 1994 and unpublished data, Bylund 1995), and trapping data of adult parasitoids in different phases of the cycle (Nuorteva and Jussila 1969), show that parasitism rates and parasitoid densities lag behind *E. autumnata* population densities (figure 8.1).

To test whether larval parasitism can produce the 9- to 10-year cycles of northern *E. autumnata* populations, we developed population models based on long-term monitorings of *E. autumnata* densities and parasitism rates. The mechanistic model was parameterized according to time series data collected in Abisko, northern Sweden (figure 8.1). We fit the data to five different models: Nicholson–Bailey, Nicholson–Bailey with host limitation in the Ricker form, Nicholson–Bailey with host limitation and parasitoid interference, Nicholson–Bailey with host limitation and type II functional response, and Nicholson–Bailey with host limitation in the theta-Ricker form, parasitoid interference, and type II functional response. The best fit (62% variance explained) was acquired with the last and most complex model (other models, 19–34% variance explained):

$$R_t = R_0\left[1 - \left(\frac{N_{t-1}}{K}\right)^\theta\right] - \frac{aP_{t-1}}{1 + ahN_{t-1} + asP_{t-1}} \quad (8.1)$$

where $R_t = \ln(N_t/N_{t-1})$ is the realized per-capita rate of increase over a year, R_0 is the maximum per-capita rate of increase, N is abundance of a host population, P is abundance of parasitoids, a is the searching rate, s is inter-

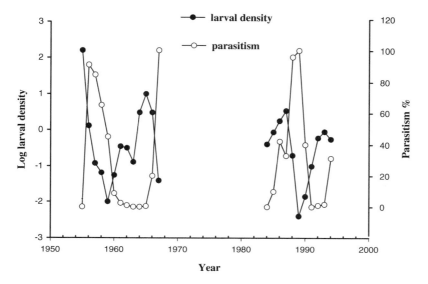

Figure 8.1 Time series of *E. autumnata* population densities and rates of parasitism in mountain birch forest of Abisko, northern Sweden (data collected by O. Tenow and H. Bylund; see Bylund 1995).

ference with other parasitoids, h is handling time, and θ determines the shape of the density dependence.

Because the estimate of h was negative (-0.03), we assumed no handling time effect and, as K was almost unity (1.13), it was also discarded. Furthermore, because θ was nearly zero (0.03), we set $(N_{t-1})^\theta = \ln N_{t-1}$ and replaced $R_0 \cdot \theta$ with g, which determines the strength of density dependence. These adjustments reduced the number of parameters without much reduction in goodness of fit (55% of the variance explained):

$$R_t = R_0 - g \log N_{t-1} - \frac{aP_{t-1}}{1 + asP_{t-1}}, \tag{8.2}$$

with the following parameter estimates (±SE): $R_0 = 0.48(\pm 1.61)$, $g = 0.62(\pm 0.22)$, $a = 202(\pm 696)$, $s = 0.31(\pm 0.18)$.

For parasitoids (all species combined), the corresponding model was

$$P_t = cN_{t-1}\left[1 - \exp\left(\frac{aP_{t-1}}{1 + asP_{t-1}}\right)\right], \tag{8.3}$$

where a and s are the same as in equation (8.2), and c is an unknown probability of surviving winter.

Simulations with the estimated parameters and a value of $c = 0.2$ for the probability of winter survival produced oscillations with a period of 7 years, somewhat shorter than the empirical cycles and with much lower amplitude. Changing the value of winter survival had minor effects on cycle period and amplitude. However, various kinds of dynamics are possible within the standard errors around parameter estimates and different winter survival values. We tested whether there were any statistically plausible combinations of parameters that would match the empirical data using the method of probes (Turchin 2001); that is, we varied the value of each parameter within its standard error and selected the parameter estimates that produced simulated dynamics most closely matching observed dynamics. Simulations with the parasitoid model using fine-tuned estimates produced cyclic dynamics with a period similar to that of northern populations (figure 8.2). Simulated peak densities averaged about 12 larvae/100 short shoots, which is also reasonable, although densities may reach 160 larvae/100 short shoots during the most severe outbreaks (Bylund 1995). This indicates that it is possible to generate realistic endogenous cycles with an empirically plausible parasitism model. We emphasize that the parasitoid function is a four-parameter nonlinear model with relatively large standard errors, which produced a very broad range of dynamics, only a subset of which match the empirical cycles.

In order to evaluate the ability of the mechanistic model to explain the observed dynamics, we also fitted a simple second-order logarithmic model to the autumnal moth data:

$$R = a + b \log N_{t-1} + c \log N_{t-2}, \tag{8.4}$$

which explained 53% of the variation with parameters ($\pm$SE) $a = -0.90(\pm 0.39)$, $b = -0.097(\pm 0.20)$, $c = -0.55(\pm 0.17)$. Thus, the mechanistic parasitoid–host model fitted the data somewhat better than a purely phenomenological time series model.

There are inevitably other factors, not included in our model, that affect population dynamics of northern *E. autumnata*. This is emphasized by the relatively low fit of the parasitism model to empirical parasitism rates (maximum 55%). Furthermore, we were unable to reproduce the very high densities that characterize some outbreaks. Extreme outbreak densities may require interactions among more than one endogenous feedback process, perhaps even positive feedback, or a coincidence of the increase phase with stochastic variation that favors population growth.

8.3.3.2 Delayed Inducible Resistance

Delayed inducible resistance (DIR) of host plant has been proposed to generate cyclic density fluctuations of insect herbivores (Rhoades 1985, Underwood 1999), including *E. autumnata* (Haukioja and Hakala 1975, Haukioja 1980, Haukioja et al. 1988, Neuvonen and Haukioja 1991). Numerous studies have demonstrated delayed reduction in *E. autumnata* performance after experimental damage to mountain birch foliage (for a review, see Ruohomäki et al. 1992, 2000). Both survival and fecundity of *E. autumnata* tend to be reduced as a consequence of DIR, although the strength of DIR has been highly variable between years and experiments (Haukioja et al. 1985, Ruohomäki et al. 1992, 2000). In some experiments,

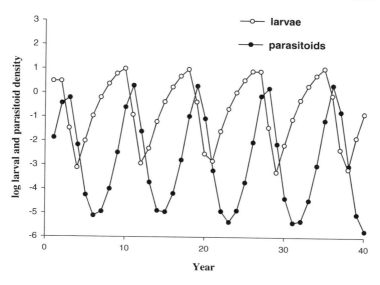

Figure 8.2 Simulation of the population dynamics of *E. autumnata* and its larval parasitoids. Parameter estimates used in this simulation were the result of a search for the parameter combination that best matched the observed dynamics within the constraint of lying within 1 SE of the initial estimates. ($a = 1.2$, $g = 0.6$, $s = 0.15$, $R_0 = 1.2$, $c = 0.2$). Simulations based on the initial estimates produced oscillations with a shorter period (7 years) and much lower amplitude.

DIR has been demonstrated to cause up to 70–80% reduction in *E. autumnata* reproductive capacity 1 year after defoliation, although in most experiments the effect has been less than 20% (Ruohomäki et al. 1992, 2000). To produce 10-year population cycles, the relaxation of DIR should take several years (May 1973, Karban and Baldwin 1997) and, in fact, the effect of DIR in mountain birch may last at least 4 years (Haukioja 1982). Thus, DIR is theoretically capable of producing the delayed negative feedback required to create population cycles of *E. autumnata*.

However, there are no time series data on the strength of DIR under natural conditions. Moreover, it may not even be possible to collect such data, as the resistance of a tree is not as simple to assess as, for instance, the parasitism rate. There is no single trait to be measured that would indicate the level of resistance. But, during the early 1990s, Kaitaniemi et al. (1999) studied the effect of DIR on *E. autumnata* populations during and after a natural outbreak in Finnish Lapland. They reared larvae on trees that had been naturally defoliated during previous years and on trees protected from defoliation by an insecticide, and trapped adult moths at sites with and without an ongoing outbreak. Although the chemical quality of birch foliage was lower the year after defoliation (Kaitaniemi et al. 1998), differences in fecundity and survival of bagged *E. autumnata* were weak or absent (Kaitaniemi et al. 1999). Thus, DIR was apparently not involved during this particular outbreak. Because defoliation rates generally remained less than 50%, lack of food could not explain the density decline either.

A problem with studying DIR is that it can operate by increasing the predation risk. That would be impossible to study by monitoring larval growth and survivorship within bag enclosures only. Kaitaniemi and Ruohomäki (2001) found that the presence of *E. autumnata* larvae on a tree significantly increased the disappearance rate of free-living larvae on the same tree the following spring. Although we have no other data to demonstrate whether this is a general trend in this system, insect feeding is known to selectively guide parasitoids to the plant (De Moraes et al. 1998).

Different factors might be important in terminating different *E. autumnata* outbreaks (Bylund 1995), as in *Zeiraphera diniana* (Lepidoptera, Tortricidae) (Baltensweiler 1993). Thus, we cannot rule out the contribution of DIR on the basis of experiments during only one outbreak, and more studies on the effects of DIR during natural *E. autumnata* outbreaks are needed. Considering the relatively weak effects of DIR, compared with the almost 100% parasitism rates, it is unlikely that DIR alone would be responsible for the cycles of *E. autumnata*.

8.3.3.3 Other Factors

All predators attacking *E. autumnata* are apparently generalists, and therefore, are not likely to give rise to delayed negative feedback or to cause *E. autumnata* cycles. However, they may modify the dynamics by causing local suppression or contribute to the termination of outbreaks.

Evidence for pathogens of *E. autumnata* is rare and appreciable mortality from diseases has been detected only during a few outbreaks (Tenow 1972, Ruohomäki personal observations). Thus, pathogens too seem unlikely to generally contribute to population cycles of *E. autumnata*.

8.4 What Causes Spatial Variation in Outbreak Severity?

During the peak density years of the *E. autumnata* cycle, outbreaks are observed in only some populations. Thus, during each outbreak period, certain forest stands are especially suitable for rapid population growth. This was also illustrated by our parasitoid model, which produces cycles with the correct period, but never especially high densities that are observed during severe outbreaks. Thus, especially suitable physical and/or biological conditions may be needed in order for populations to reach very high densities.

Temperatures below $-36°C$ kill *E. autumnata* eggs (e.g., Niemelä 1979, Nilssen and Tenow 1990) and are not uncommon in northern Fennoscandia. Thus, winter temperatures surely limit some outbreaks (e.g., Ruohomäki et al. 1997, Virtanen et al. 1998). In Lapland, winter temperatures are particularly severe at low altitudes, and this is probably why outbreaks are usually observed at high altitudes (Tenow 1972, Ruohomäki et al. 1997). Rapid density increase may therefore be possible only in areas with mild winters and resulting low egg mortality.

E. autumnata outbreaks mostly take place in old mountain birch forests (Tenow and Bylund 1989, Bylund 1995, 1997, Ruohomäki et al. 1997). Higher foliage quality, lower number of parasitoids, and protected oviposition sites due to large amounts of lichen on trees (Bylund 1997) have been proposed to explain the better suitability of old forests for *E. autumnata*. However, empirical evidence does not support the importance of any of these factors (Tammaru et al. 1995, Bylund 1997, Ruohomäki et al. 2000). Thus, the causality behind the apparently better suitability of old mountain birch forests remains unsolved.

On a smaller scale, predation by ants (*Formica* spp.) creates very localized "green islands" around ant mounds (Laine and Niemelä 1980, Niemelä and Laine 1986, Karhu and Neuvonen 1998). Indeed, ants may be important in the suppression of local infestations and affect the fine scale spatial distribution of outbreaks.

Factors involving positive feedback may increase the population growth rate. One such potential factor is herbivore-triggered host plant susceptibility (Haukioja et al. 1990, Karban and Thaler 1999). During high natural densities of *E. autumnata*, most feeding occurs on the apical shoots. Therefore, the breaking of apical dominance may result in better foliage quality for the same or the next generation(s) (Haukioja et al. 1990), potentially resulting in a higher population growth rate. However, increased susceptibility of defoliated foliage was not detected in experimental studies or during a natural outbreak in the 1990s (Kaitaniemi et al. 1997).

One or several of the factors mentioned above, or other unknown factors, are likely involved in the spatial distribution of outbreaks. We want to stress the importance of determining not only the causes of cyclicity, but also the reasons for the very rapid increases of some populations.

8.5 What Prevents Outbreaks in Southern Populations?

Genetic differences in body size and fecundity cannot explain the absence of outbreaks in southern populations of *E. autumnata* (Haukioja et al. 1988, Ruohomäki and Haukioja 1992a) and, thus, the strikingly different dynamics between outbreaking northern populations and stable southern populations must be explained by other causes.

In general, *E. autumnata* larvae grow larger in the south than in the north (Ruohomäki and Haukioja 1992a), and this seems to falsify the idea that southern populations do not reach outbreak levels because of higher constitutive resistance in southern birches. A further trait related to plants that might contribute to the lack of outbreaks in southern Fennoscandia is the higher plant species diversity of forests in southern Fennoscandia. But, if this was important, we would still expect local outbreaks in pure birch stands in the south, and this does not occur.

In southern populations, pupal predation by small mammalian predators is temporally density-dependent without a lag (Tanhuanpää et al. 1999). Thus, pupal predators may regulate and stabilize southern populations. Pupal predation has been implicated in the regulation of many other Lepidoptera (East 1974, Bechinski and Pedigo 1983, Bauer 1985, Walsh 1990, Weseloh 1990, Cook et al. 1994, Pearsall and Walde 1994). However, small mammals may be generally unable to respond to very high prey densities (Hanski 1992). This was supported by the observance of density-independent pupal predation during an outbreak of *E. autumnata* in Lapland (Tanhuanpää et al. 1999). Thus, suppression by other mortality factors may be crucial if small mammalian pupal predators are to regulate autumnal moth populations. For example, in the winter moth (*Operophtera brumata*: Lepidoptera, Geometridae), larval parasitoids suppress host numbers to a level at which regulation by pupal predators is possible (Roland 1994). In nonoutbreaking *E. autumnata* populations, 95% of larvae do not survive through the larval and pupal stage. Most of the mortality takes place during the larval stage and is due to parasitism and avian predators (Tanhuanpää et al. 2001). Further, parasitism rates remain relatively constant at about 30% (Teder et al. 2000). Therefore, the fact that larval parasitoids are generalists in the south and functional specialists in the north may be crucial to the contrasting dynamics of *E. autumnata* populations in southern and northern Finland.

8.6 Conclusions

Northern populations of *E. autumnata* exhibit cyclic population dynamics, which are most easily understood as the result of delayed density dependence. Two specific ecological processes known to produce delayed density dependence in *E. autumnata* are DIR of the host plant and parasitism by specialist wasps. We used empirically based population modeling to test whether parasitism is, by itself, able to produce population cycles of *E. autumnata*. The model was parameterized with empirical data. Simulations with the initial parameters failed to produce dynamics that matched empirical time series; that is, the period and amplitude were much lower than the real oscillation. However, it was possible to find statistically plausible parameters that could produce realistic population dynamics. So, larval parasitism by itself can explain the endogenous population cycles of northern *E. autumnata*. However, this conclusion must be qualified by noting that many different dynamics can be produced with statistically plausible parameters (and many of them do not match the data). Expansion of the time series data for this host–parasitoid system should allow refinement of the endogenous dynamics expected from larval parasitism alone. This will also facilitate tests of whether the multiple density-dependent feedbacks in this system interact to produce endogenous dynamics that cannot be easily predicted from the characteristics of any one ecological feedback process.

REFERENCES

Baltensweiler, W. 1993. Why the larch bud-moth cycle collapsed in the subalpine larch–cembran pine forests in the year 1990 for the first time since 1850. *Oecologia* 94: 62–66.

Bauer, G. 1985. Population ecology of *Pardia tripunctana* Schiff. & *Notocelia roborana* Den. & Schiff. (Lepidoptera, Tortricidae)—an example of 'Equilibrium species.' *Oecologia* 65: 437–441.

Bechinski, E. J. and L. P. Pedigo. 1983. Green cloverworm (Lepidoptera: Noctuidae) population dynamics: pupal life table studies in Iowa soybeans. *Environ. Entomol.* 12: 656–661.

Berryman, A. A. 1981. *Population systems: a general introduction.* Plenum Press, New York.

Berryman, A. A. 1996. What causes population cycles of forest Lepidoptera? *Trends Ecol. Evol.* 11: 28–32.

Bylund, H. 1995. *Long-term interactions between the autumnal moth and mountain birch: the roles of resources, competitors, natural enemies and weather.* Ph.D. thesis, Swedish University of Agricultural Sciences, Uppsala, Sweden.

Bylund, H. 1997. Stand age-structure in a low population peak of *Epirrita autumnata* in mountain birch forest. *Ecography* 20: 319–326.

Cook, S. P., F. P. Hain, and H. R. Smith. 1994. Oviposition and pupal survival of gypsy moth (Lepidoptera; Lymantriidae) in Virginia and North Carolina pine–hardwood forests. *Environ. Entomol.* 23: 360–366.

De Moraes, C. M., W. J. Lewis, P. W. Pare, H. T. Alborn, and J. H. Tumlinson. 1998. Herbivore-infested plants selectively attract parasitoids. *Nature* 393: 570–573.

East, R. 1974. Predation on the soil-dwelling stages of the winter moth at Wytham woods. *J. Anim. Ecol.* 43: 611–626.

Faeth, S. H. 1987. Community structure and folivorous insect outbreaks: the roles of vertical and horizontal interactions. In P. Barbosa and J. C. Schultz (Eds.) *Insect outbreaks.* Academic Press, New York, pp. 135–171.

Hanski, I. 1990. Density-dependence, regulation and variability in animal populations. *Philos. Trans. Roy. Soc. Lond., Ser. B* 330: 141–150.

Hanski, I. 1992. Insectivorous mammals. In H. J. Crawley (Ed.) *Natural enemies. The population biology of predators, parasites and diseases.* Blackwell Scientific, Oxford, pp. 163–187.

Haukioja, E. 1980. On the role of plant defenses in the fluctuations of herbivore populations. *Oikos* 35: 202–213.

Haukioja, E. 1982. Inducible defences of white birch to a geometrid defoliator, *Epirrita autumnata.* Proceedings of the 5th International Symposium on Insect–Plant Relationships, Wageningen, 1982.

Haukioja, E. and T. Hakala. 1975. Herbivore cycles and periodic outbreaks. Formulation of general hypothesis. *Rep. Kevo Subarct. Res. Stat.* 12: 1–9.

Haukioja, E., J. Suomela, and S. Neuvonen. 1985. Long-term inducible resistance in birch foliage: triggering cues and efficacy of a defoliator. *Oecologia* 65: 363–369.

Haukioja, E., S. Neuvonen, S. Hanhimäki, and P. Niemelä. 1988. The autumnal moth in Fennoscandia. In A. A. Berryman (Ed.) *Dynamics of forest insect populations. Patterns, causes, implications.* Plenum Press, New York, pp. 163–178.

Haukioja, E., K. Ruohomäki, J. Senn, J. Suomela, and M. Walls. 1990. Consequences of herbivory in the mountain birch (*Betula pubescens* ssp. *tortuosa*): importance of the functional organization of the tree. *Oecologia* 82: 238–247.

Hunter, A. F. 1991. Traits that distinguish outbreaking and nonoutbreaking Macrolepidoptera feeding on northern hardwood trees. *Oikos* 60: 275–282.

Hunter, A. F. 1995. Ecology, life history, and phylogeny of outbreak and nonoutbreak species. In N. Cappuccino and P. W. Price (Eds.) *Population regulation. New approaches and synthesis.* Academic Press, New York, pp. 41–64.

Kaitaniemi, P. and K. Ruohomäki. 2001. Sources of variability in plant resistance against insects: free caterpillars show strongest effects. *Oikos* 95: 461–470.

Kaitaniemi, P., K. Ruohomäki, and E. Haukioja. 1997. Consumption of apical buds as a mechanism of alleviating host plant resistance for *Epirrita autumnata* larvae. *Oikos* 78: 230–238.

Kaitaniemi, P., K. Ruohomäki, V. Ossipov, E. Haukioja, and K. Pihlaja. 1998. Delayed induced changes in the biochemical composition of host plant leaves during an insect outbreak. *Oecologia* 116: 182–190.

Kaitaniemi, P., K. Ruohomäki, T. Tammaru, and E. Haukioja. 1999. Induced resistance of host tree foliage during and after a natural insect outbreak. *J. Anim. Ecol.* 68: 382–389.

Karban, R. and I. T. Baldwin. 1997. *Induced responses to herbivory.* University of Chicago Press, Chicago, Ill.

Karban, R. and J. S. Thaler. 1999. Plant phase change and resistance to herbivory. *Ecology* 80: 510–517.

Karhu, K. J. and S. Neuvonen. 1998. Wood ants and a geometrid defoliator of birch: predation outweights beneficial effects through the host plant. *Oecologia* 113: 509–516.

Laine, K. J. and P. Niemelä. 1980. The influence of ants on the survival of mountain birches during an *Oporinia autumnata* (Lep., Geometridae) outbreak. *Oecologia* 47: 39–42.

Lehtonen, J. and R. K. Heikkinen. 1995. On the recovery of mountain birch after *Epirrita* damage in Finnish Lapland, with a particular emphasis on reindeer grazing. *Ecoscience* 2: 349–356.

Leirs, H., N. C. Stenseth, J. D. Nichols, J. E. Hines, R. Verhagen, and W. Verheyen. 1997. Stochastic seasonality and nonlinear density-dependent factors regulate population size in an African rodent. *Nature* 389: 176–180.

Lindström, J., H. Kokko, and E. Ranta. 1996. There is nothing new under the sunspots. *Oikos* 77: 565–568.

Mason, R. R. 1987. Nonoutbreak species of forest lepidoptera. In P. Barbosa and J. C. Schultz (Eds.) *Insect outbreaks.* Academic Press, New York, pp. 31–57.

May, R. M. 1973. *Stability and complexity in model ecosystems.* Princeton University Press, Princeton, N.J.

Myers, J. H. and L. D. Rothman. 1995. Field experiments to study regulation of fluctuating populations. In N. Cappuccino and P. W. Price (Eds.) *Population regulation. New approaches and synthesis.* Academic Press, New York, pp. 229–250.

Neuvonen, S. and E. Haukioja. 1991. The effects of inducable resistance in host foliage on birch feeding herbivores. In D. W. Tallamy and M. J. Raupp

(Eds.) *Phytochemical induction by herbivores.* John Wiley, New York, pp. 277–291.

Niemelä, P. 1979. Topographical delimitation of *Oporinia*-damages: experimental evidence of the effect of winter temperature. *Rep. Kevo Subarct. Res. Stat.* 15: 33–36.

Niemelä, P. and K. J. Laine. 1986. Green islands—predation not nutrition. *Oecologia* 68: 476–478.

Nilssen, A. and O. Tenow. 1990. Diapause, embryo growth and supercooling capacity of *Epirrita autumnata* eggs from northern Fennoscandia. *Entomol. Exp. Appl.* 57: 39–55.

Nuorteva, P. and R. Jussila. 1969. Incidence of ichneumonids on a subarctic fell after a calamity of the moth *Oporinia autumnata* (Bkh.) on birches. *Ann. Entomol. Fenn.* 37: 96.

Pearsall, I. A. and S. J. Walde. 1994. Parasitism and predation as agents of mortality of winter moth populations in neglected apple orchards in Nova Scotia. *Ecol. Entomol.* 19: 190–198.

Rhoades, D. F. 1985. Offensive–defensive interactions between herbivores and plants: their relevance in herbivore population dynamics and ecological theory. *Am. Nat.* 125: 205–238.

Roland, J. 1994. After a decline: what maintains low winter moth density after successful biological control? *J. Anim. Ecol.* 63: 392–398.

Ruohomäki, K. 1992. Wing size variation in *Epirrita autumnata* (Lep., Geometridae) in relation to larval density. *Oikos* 63: 260–266.

Ruohomäki, K. 1994. Larval parasitism in outbreaking and non-outbreaking populations of *Epirrita autumnata* (Lepidoptera, Geometridae). *Entomol. Fenn.* 5: 27–34.

Ruohomäki, K. and E. Haukioja. 1992a. Interpopulation differences in pupal size and fecundity are not associated with occurrence of outbreaks in *Epirrita autumnata* (Lepidoptera, Geometridae). *Ecol. Entomol.* 17: 69–75.

Ruohomäki, K. and E. Haukioja. 1992b. No evidence of genetic specialization to different natural host plants within or among populations of a polyphagous geometrid moth *Epirrita autumnata*. *Oikos* 63: 267–272.

Ruohomäki, K., S. Hanhimäki, E. Haukioja, L. Iso-Iivari, S. Neuvonen, P. Niemelä, and J. Suomela. 1992. Variability in the efficacy of delayed inducible resistance in mountain birch. *Entomol. Exp. Appl.* 62: 107–115.

Ruohomäki, K., T. Virtanen, P. Kaitaniemi, and T. Tammaru. 1997. Old mountain birches at high altitudes are prone to outbreaks of *Epirrita autumnata* (Lepidoptera: Geometridae). *Environ. Entomol.* 26: 1096–1104.

Ruohomäki, K., M. Tanhuanpää, M. P. Ayres, P. Kaitaniemi, T. Tammaru, and E. Haukioja. 2000. Causes of cyclicity of *Epirrita autumnata* (Lepidoptera, Geometridae)—grandiose theory and tedious practice. *Popul. Ecol.* 42: 211–223.

Seppänen, E. J. 1970. Suurperhostoukkien ravintokasvit (The food-plants of the larvae of the Macrolepidoptera of Finland). In *Animalia Fennica 14*. Werner Söderström, Porvoo, Helsinki.

Sinclair, A. R. E. and R. P. Pech. 1996. Density dependence, stochasticity, compensation and predator regulation. *Oikos* 75: 164–173.

Speight, M. R., M. D. Hunter, and A. D. Watt. 1999. *Ecology of insects. Concepts and applications.* Blackwell Scientific, Oxford.

Tammaru, T. 1998. Determination of adult size in a folivorous moth: constraints at instar level? *Ecol. Entomol.* 23: 80–89.

Tammaru, T. and E. Haukioja. 1996. Capital breeders and income breeders among Lepidoptera—consequences to population dynamics. *Oikos* 77: 561–564.

Tammaru, T., P. Kaitaniemi, and K. Ruohomäki. 1995. Oviposition choices of *Epirrita autumnata* (Lepidoptera: Geometridae) in relation to its eruptive population dynamics. *Oikos* 74: 296–304.

Tammaru, T., P. Kaitaniemi, and K. Ruohomäki. 1996. Realized fecundity in *Epirrita autumnata* (Lepidoptera: Geometridae): relation to body size and consequences to population dynamics. *Oikos* 77: 407–416.

Tanhuanpää, M., K. Ruohomäki, P. Kaitaniemi, and T. Klemola. 1999. Different impact of pupal predation on populations of *Epirrita autumnata* (Lepidoptera; Geometridae) within and outside the outbreak range. *J. Anim. Ecol.* 68: 562–570.

Tanhuanpää, M., K. Ruohomäki, and E. Uusipaikka. 2001. High larval predation rate in nonoutbreaking populations of a geometrid moth. *Ecology* 82: 281–289.

Teder, T., M. Tanhuanpää, K. Ruohomäki, P. Kaitaniemi, and J. Henriksson. 2000. Temporal and spatial variation of larval parasitism in non-outbreaking populations of a folivorous moth. *Oecologia* 123: 516–524.

Tenow, O. 1972. *The outbreaks of* Oporinia autumnata *Bkh. &* Operophtera *spp. (Lep. Geometridae) in the Scandinavian mountain chain and northern Finland 1862–1968*. Ph.D. thesis, University of Uppsala, Uppsala, Sweden.

Tenow, O. and H. Bylund. 1989. A survey of winter cold in the mountain birch/ *Epirrita autumnata* system. *Mem. Soc. Fauna Flora Fenn.* 65: 67–72.

Turchin, P. 1990. Rarity of density dependence or population regulation with lags? *Nature* 344: 660–663.

Turchin, P. 2002. *Complex population dynamics: a theoretical/empirical synthesis*. Princeton University Press, Princeton, N.J. (in press).

Turchin, P. and A. Berryman. 2000. Detecting cycles and delayed density-dependence: a comment on Hunter and Price (1998). *Ecol. Entomol.* 25: 119–121.

Underwood, N. 1999. The influence of plant and herbivore characteristics on the interaction between induced resistance and herbivore population dynamics. *Am. Nat.* 153: 282–294.

Virtanen, T., S. Neuvonen, and A. Nikula. 1998. Modeling topoclimatic patterns of egg mortality of *Epirrita autumnata* (Lepidoptera: Geometridae) with a Geographical Information System: predictions for current climate and warmer climate scenarios. *J. Appl. Ecol.* 35: 311–322.

Waldmeier, M. 1961. *The sunspot activity in the years 1610–1960*. Schultess, Zurich.

Wallner, W. E. 1987. Factors affecting insect population dynamics: differences between outbreak and non-outbreak species. *Ann. Rev. Entomol.* 32: 317–340.

Walsh, P. J. 1990. Site factors, predators and pine beauty moth mortality. In A. D. Watt, S. R. Leather, M. D. Hunter, and N. A. Kidd (Eds.) *Population dynamics of forest insects*. Intercept, Andover, UK, pp. 245–252.

Weseloh, R. M. 1990. Gypsy moth predators: an example of generalist and specialist natural enemies. In A. D. Watt, S. R. Leather, M. D. Hunter and N. A. Kidd (Eds.) *Population dynamics of forest insects*. Intercept, Andover, UK, pp. 233–243.

9

Population Cycles

Inferences from Experimental, Modeling, and Time Series Approaches

Xavier Lambin, Charles J. Krebs, Robert Moss, and Nigel G. Yoccoz

9.1 Introduction

Some of the most interesting debates in population ecology have taken place within the context of population cycles. Their study has been a fertile ground for the development of ideas on how population models should be formulated and confronted with data. It is the setting in which the use of field experiments became established in ecology (e.g., Krebs and DeLong 1965), and also the context of many methodological and conceptual developments in the fields of population demography (Leslie and Ranson 1940), pest management (Berryman 1982), and community dynamics (Sinclair et al. 2000). Yet, as with many other issues in population dynamics, identifying without ambiguity the causes of population cycles in general, and for any organism in particular, continues to prove an extraordinarily difficult task.

The major purpose of this book is to review recent research developments on the role of food web architecture, and more specifically on the effects of food, predators, and pathogens in population cycles. Its stated aim is to present evidence that population cycles could be caused by food web architecture in some natural systems. Whereas in chapter 1 Alan Berryman promotes a research program centered on the analysis of time series data for formulating, selecting, and even testing hypotheses on population cycles, the case studies encompass a much broader diversity of research approaches. The authors and coworkers of the seven case studies have combined time series analysis, model building, natural history observation, and experiments in different proportions to reach the conclusion that trophic interactions play an important role in generating cyclic dynamics. This diversity of approaches reflects, in part, a taxonomic divide between vertebrates and invertebrates, experiments being more common with the former, but also profound differences in research traditions. Indeed, the investment required to estimate population size and quantify the causes of mortality of moths and beetles is substantially less than that required for estimating the abundance of voles, hares, and grouse and their predators. From these practical constraints, divergent research traditions have evolved. Despite reaching a broad consensus that trophic interactions are "important" to population cycles of diverse taxonomic groups, the varying strength of the conclusions reached betrays profound disagreements on what inferences ecologists should be willing to make from different strands of data. The statement that trophic interactions are "important" has the benefit of being uncontroversial, but cannot be defined operationally. A key contentious issue, which forms the thread of this chapter, is whether experiments are necessary for separating correlation from causation or whether time series analysis provides an alternative way to eliminate false hypotheses and, given appropriate data, eventually identify the cause of population cycles.

In this chapter we critically review the findings of the empirical chapters, while focusing on the contribution of the research approaches being used. We accept the conclusion that every chapter of this book presents convincing evidence that trophic interactions *could* cause population cycles in the species and populations under study. However, we hold the view that, even in the absence of a well-formulated competing hypothesis, scientific rigor demands that we subject our hypotheses to critical testing, whether through experiment, deductive logic, or other means. Failure to do this has led to the multiplication of poorly scrutinized hypotheses or to the acceptance of untested explanations. Throughout our discussion, we highlight the various stages at which hypotheses erected as explanations for population cycles should be subjected to attempted falsification.

Our discussion is divided into three sections. First, we distinguish between different types of questions in need of answering. Second, we consider the strength and limitations of approaches based on time series, and the contribution of modeling to the formalization and testing of hypotheses on the cause of population cycles. Finally, we review the contribution that experiments have made and should make to subjecting hypotheses dealing with the causes of population cycles to rigorous testing.

9.2 Cyclical Population Dynamics and the Dynamics of Cyclical Populations

The primary question that contributors to this book aim to answer is: What is (are) the specific causal factor(s) responsible for the oscillatory nature of cyclic population fluctuations? Many species have populations with both cyclical and noncyclical populations, thus species-specific features are unlikely to hold the answer. In contrast, features of ecosystems could differentiate populations with cyclical oscillations from those whose numbers fluctuate, sometimes violently, but without indication of a regular period. Thus trophic interactions form a plausible answer. A second, related question seeks to determine: What processes explain spatial and temporal variation in the dynamics of cyclical populations? Populations of several species display cycles over large geographical areas, but amplitudes and periods often vary markedly over such areas. Whether the process accounting for the oscillatory nature of the dynamics also accounts for the spatial variation in dynamics, or whether two or more processes interact in shaping the dynamics observed in a given region, is an important question.

Successive fluctuations of the same population in the same location may differ in amplitude and in the abundance of natural enemies (figure 9.1). Populations of the autumnal moth and small rodents in Fennoscandia offer striking examples. Oscillations of autumnal moths in Lapland sometimes reach outbreak densities with associated widespread defoliation and tree death at a given site. Subsequent cycles at the same location may be of such low amplitude as to be nearly undetectable (chapter 8). The fluctuations of small rodent populations also vary in amplitude along geographical gradients and over time. Up to five small rodent species in Finnish Lapland experienced high amplitude, largely synchronous cycles with low trough densities from 1950 to 1985. More recently, and at the same sites, *Microtus* voles that were previously abundant have become scarce and *Clethrionomys* voles now experience much lower amplitude, yet still distinctly cyclical oscillations (figure 9.1a).

Oscillations that vary in amplitude or period, but nevertheless remain distinctly cyclic, suggest that factors responsible for variation in amplitude or period are not necessary to explain the cyclical nature of dynamics, even though they influence the dynamics. For instance, changes in the prevalence of parasitism in a cyclic host population regulated by a predator–prey interaction might result in different dynamics than would occur if only the predator–prey interactions shaped its dynamics. This could take the form of steeper, shorter population declines if epidemics occurred following high-density years, even if the parasite–host interaction itself did not cause the cycles. A related view is that specialist mammalian predators might modify the shape of cyclic population trajectories by deepening or prolonging the low phase without causing their cycles (Pearson 1966, MacLean et al. 1974, Fitzgerald 1977), a view consistent with recent experimental data with cyclic small rodents (Korpimäki and Norrdahl 1998). This could occur, for example, if predators were subject to density-dependent mortality that precluded them from showing a numerical response. A further example of such a potentially amplifying process might be found in red grouse populations in Britain. Low amplitude, long (6–11 years) and symmetrical (equal time for up and down) cycles of red grouse have been documented in northeast Scotland where average rainfall is low (e.g., Moss and Watson 2001) and, presumably, parasite transmission is low and variable because free-living stages of the parasite require high humidity to survive (Hudson 1986b, Moss et al. 1993,

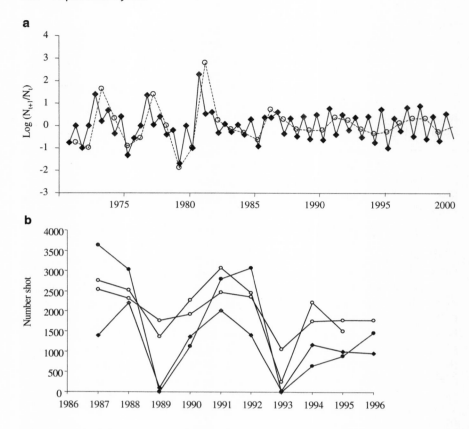

Figure 9.1 Two examples of cyclic populations with changing amplitude, suggesting that separate processes simultaneously determine the amplitude and the cyclical nature of fluctuations. (a) Changes in per-capita growth rates of bank voles (*C. glareolus*) at Pallasjärvi, Finnish Lapland remain distinctly oscillatory despite a marked change in the seasonal dynamics and amplitude of oscillations circa 1985 when, for unknown reasons, *Microtus* voles, the primary prey of least weasels, became scarce (Henttonen 2000). White circles are spring surveys, black diamonds are late summer surveys. (b) Number of red grouse (*L. lagopus scoticus*) shot on two British moors where parasite burden was reduced experimentally in 1989 and 1993 (white circles) and on two control moors (black circles) (Hudson et al. 1998). Although no attempt was made to shoot birds on the control areas in 1989 and 1993, cycle amplitude was apparently lower on the treated area relative to controls, but trajectories remained cyclical in all areas.

Saunders et al. 1999). Thus, it seems unlikely that parasites could be responsible for the long, low-amplitude cycles in the north (Moss et al. 1993, 1996). Further south, where rainfall is substantially higher, and parasite transmission presumably higher, shorter (3–5 years), high-amplitude cycles are asymmetrical (slow up, fast down) because precipitous declines are associated with a high prevalence of cecal threadworms (Hudson et al. 1992). There are three possible explanations for this observed pattern. (1) Cycles are caused by a parasite–host interaction but the period and amplitude are affected by variations in rainfall (or some other related variable). (2) Cycles are caused by some other delayed density-dependent process (e.g., kin selection) but the amplitude and period are affected by parasitism in wetter regions and/or times; for example, in areas of low average rainfall, parasite outbreaks would tend to occur when wet summers coincide with high grouse density, resulting in precipitous declines in the grouse population, an increase in the amplitude and, possibly, a shortening of the period of the cycle. (3) Cycles in the north are caused by a different mechanism than in the south, which may be the result of host–parasitoid interactions; for example, different feedback mechanisms dominate the dynamics in the north (chapter 1, Berryman 1993). An experiment capable of differentiating between these hypotheses has yet to be performed.

We see two main reasons for stressing the distinction between factors influencing the dynamics of cyclical populations and those responsible for the cyclical nature of their dynamics. For brevity we focus on amplitude, while noting that some of the same considerations apply to period. First, focusing on the capacity of causal processes to account for observed variations in population growth rates, instead of on those responsible for the cyclical nature of dynamics, rules out a priori a large number of processes that may nevertheless be causally involved. Variations in life history and demographic traits not caused by resources or natural enemies are almost always subtle and hence, by definition, unlikely to be the sole cause of wide amplitude fluctuations such as those observed pre-1982 in bank voles in Lapland (figure 9.1a). Yet, variations in those subtle factors, rather than more potent agents of mortality, may distinguish cyclical populations from those which experience erratic fluctuations (e.g., Oli and Dobson 2001). Second, given geographical variability in cycle amplitude, even between populations of the same species, restricting the scope of specific causal hypotheses to cycles with similar amplitude precludes much scope for refutation. The assumption that cycles with different amplitude result from different causal processes would automatically lead to an accumulation of explanations that could not be refuted by data.

9.3 Testability and Paradigms in the Study of Population Cycles

Two highly contrasting paradigms have been invoked when tackling these problems, and they call for different research agendas. One paradigm is summarized by Berryman (chapter 1). Briefly, Berryman first distinguishes between cycles that result from unidirectional causal processes (*exogenous factors*) and those driven by delayed negative feedback loops (*endogenous factors*). Determining the causes of endogenous cycles involves the principle of feedback dominance (limiting factors), which implies that only one of many possible delayed feedback loops will dominate the dynamics of a species at one time and place (Berryman 1993). It therefore follows that the research challenge is to distinguish *dominant* feedback loops from those that are *subsidiary*. One way to achieve this using time series is to determine the proportion of the variation in the realized per-capita growth rate of a population that can be statistically explained by a time series of another species. If one feedback loop is broken, another may take effect. Accordingly, blocking the causal feedback may change the nature of the cycle and the causal process but not necessarily stop the cycles. This paradigm calls for time series of abundance of as many species as possible, for these are required to establish the strength of the feedback, as determined by correlation analysis, between the various components of the web. The strongest negative feedback loop is considered to dominate the dynamics and, thereby, is most likely to be responsible for the observed population cycles. Feedback loops that account for a lower proportion of the variance in the realized per-capita growth rate are deemed *subsidiary*. The capacity of models of trophic interaction parameterized from time series to reproduce the observed population trajectory is then compared with the output of a model based on the explanatory variable that showed the strongest correlation with the growth rate of the focal population. In this context, the role of experiments is largely restricted to establishing the existence of feedback loops and testing the predictions of feedback models. According to the feedback dominance paradigm, experiments aimed at stopping the cycle may be doomed to failure because removing a dominant feedback process may lead to a previously subsidiary loop dominating the system.

An alternative paradigm, embraced by many experimentalists, has been to search for conditions that are *necessary* or *sufficient* for a given population or group of populations to show regular cycles. A necessary condition must always be present for an effect to occur, but may not be sufficient to cause the effect. A cause may be sufficient to result in an effect, but if the same effect occurs in its absence, the cause is not necessary (Chitty 1960, Moss and Watson 2001). Accordingly, a process that accounts for a large proportion of the variation in a population's growth rate need not be responsible for the cyclical nature of its dynamics if, in its absence, the population nevertheless experiences cyclical oscillations, albeit with a lower amplitude. Thus, the approach is based on refutation of hypotheses and centers on explaining the cyclical nature of dynamics instead of the overall variance in population growth rates. An assumption underlying the search for causes necessary and sufficient for cyclicity is that several processes may act, additively or not, in shaping observed trajectories. Some processes not responsible for the cyclical nature of a population trajectory may amplify, dampen, or otherwise modify oscillations caused by the necessary and sufficient processes. The challenge is to experimentally identify the subset of processes contributing to the dynamics of a population that are actually responsible for its cyclical nature.

If one subscribes to Berryman's (chapter 1) argument that delayed density dependence is the only necessary condition for generating endogenous cycles, the search is then for the specific process necessary to generate delayed density-dependent feedback in given cyclical populations. Beyond the fact that it is more satisfying to identify an ecological process rather than to detect the signature of delayed density dependence, searching for necessary and sufficient conditions provides an obvious way to generate predictions that can be tested by experiments, observational data, and time series. Assume, for example, a predator–prey cycle characterized by a delayed negative feedback loop. If experimentally preventing the delayed numerical response of a specialized predator succeeded in preventing regular cycles from taking place, then the specific endogenous feedback loop caused by predation would have been shown to be a necessary part of the explanation for the observed cycles. The test would be even more convincing if reinserting the predator's numerical response into the system was followed by a resumption of cycles. Conversely, if removing or preventing the numerical response of a predator did not substantially alter the dynamics of the prey, or if the resulting prey dynamics remained cyclical, although with a reduced amplitude, then predation would be viewed as not necessary for the cyclical dynamics of that population at that time. In the case of the thought experiment above, this outcome would not imply that predation does not contribute to the observed dynamics, but instead would indicate that some other process is responsible for the cyclical nature of the dynamics.

Identifying which process(es) are necessary for cyclical dynamics and associated density dependence need not be restricted to data obtained by manipulative experiments. Indeed, not all variables invoked as part of hypothetical explanations of cycles are amenable to experimentation. This does not preclude the application of a Popperian (refutation) approach. Under some circumstances, observational data may allow for strong inference on the cause of population cycles. If hypotheses are to remain testable in the absence of manipulative experiments, it is essential that their predictions should be spelled out quantitatively, and sufficiently diverse predictions should be generated, such that they can be confronted with empirical evidence. Constructing and parameterizing mathematical models is one way to meet these requirements. Berryman (chapter 1) discusses some of the unresolved controversies on the most appropriate formulations. Note that if specific model predictions are tested, the chosen formulation used to capture the process invoked, as well as the parameter values chosen or estimated, then become part of a specific hypothesis refutable by any inconsistent data. Another requirement for maintaining the testability of hypotheses in the absence of experiments is that their spatial and taxonomic scope should be spelled out explicitly. Many species with cyclic population dynamics have broad geographical ranges and researchers collect data and perform experiments in different locations. There is, of course, no need to presume that the same process is responsible for delayed density dependence in demographic parameters throughout the range of cyclical dynamics, although some prefer this as a starting assumption (Krebs 1996, Moss and Watson 2001). However, unless the spatial scope of a hypothesis is spelled out, it is impossible to test it.

Where biogeographical gradients are present, such as in red grouse, voles, and autumnal moths (Hanski et al. 1991, Hudson 1992, Bjørnstad et al. 1998a), parsimony dictates that the same process should be invoked to account for the cyclical nature of the dynamics in adjacent populations. A second process, or a change in the parameters of the same process, must therefore be invoked to account for the geographical variation in dynamics. For example, red grouse populations experience population cycles of varying amplitude and period ranging from 4 to 11 years in a 300 km region from northern England to northern Scotland (Williams 1985). The important question is whether the observed increase in period is caused by a change in the mechanism responsible for the cycle, or by a change in the modifying factors. Moss et al. (1993) found that worm burdens in the north were too low to generate cyclic dynamics in a parameterized parasite–host model, implying that parasites are a modifying factor of a cycle caused by some other process, or that different processes are responsible for the cycle in the north and south.

When mathematical models are parameterized using data from manipulative experiments, or gleaned from appropriate studies performed in a given geographical area, such as those of chapters 3, 5, and 8, a minimum requirement is that model predictions, and hence the hypothesis as captured by the model, should be falsifiable by any contrary data from the geographical area used for parameterizing the model. For instance, models of the specialist predation hypothesis for small mammal cycles include weasel growth parameters derived from studies conducted in Britain and Fennoscandia (chapter 3, Turchin and Hanski 1997). They must therefore be consistent with all empirical patterns in those two geographical areas. On the other hand, observations such as that of lemming cycles on Wrangel Island in Siberia, in the complete absence of any mustelid predators (Litvin and Ovsyanikov 1990), does not refute the specialist predation hypothesis as spelled out for Fennoscandia. Instead, it demonstrates that predation by mustelids is not a necessary condition for the occurrence of 3–4-year cycles in small rodents, and suggests that there may be many sufficient but no geographically widespread necessary conditions for cycles in small rodents.

Finally, care is needed when assessing the degree of support from non-experimental data for hypotheses invoking more than one process such as, for instance, specialist and generalist predators in causing and dampening population cycles in Fennoscandia, respectively. In chapter 3, Hanski and Henttonen describe six features of rodent dynamics that may be explained by predation. These include the presence of a geographical gradient in cyclicity, interspecific and spatial synchrony of fluctuations, and the recent reduction in the amplitude of fluctuations. Although the ability to explain so many aspects of the dynamics is impressive, the fact that predation could explain these six aspects of rodent dynamics does not provide support for or against the hypothesis that predation by specialists is the necessary process for causing population cycles per se in Fennoscandia. Indeed, if a gradient in generalist predation can dampen cycles caused by specialist predators, then the same gradient in generalists should be able to dampen cycles caused by other processes. The converse is of course true. The observation that cycles occur in vole populations subjected to generalist predation pressure much higher than that predicted by the model of Turchin and Hanski (1997), in which a gradient in generalist predation added to the interaction between least weasels (specialist) and voles reproduces the Fennoscandian gradient in rodent cyclicity, is a challenge to the generalist component of the hypothesis and not its specialist component (Lambin et al. 2000). Note, however, that the most productive function of a hypothesis formalized in a model is to derive new predictions and not simply account for known observations.

9.4 Time Series Analysis as a Tool for Formulating, Selecting, and Testing Hypotheses

Three case studies in this book (chapters 2, 5, and 7) illustrate well the potential of multispecies time series for developing hypotheses when appropriate data are available (see also Royama 1992, Kendall et al. 1999). When multispecies time series are available, attempts can be made to identify the structure in the food web that could create the observed dynamics, and Berryman (chapter 1) advocates the use of the R-function to this effect. For instance, the application of time series analysis, in the manner advocated by Berryman, to Münster-Swendsen's needleminer data was clearly more effective than k-factor analysis, and pointed toward the existence of a hitherto unidentified factor associated with parasitism and influencing miner fecundity. The hypothesis generated from this particular time series analysis (TSA) led to new experiments being performed and to the eventual demonstration of the existence of pseudoparasitism. Stenseth et al. (1998) applied empirical, statistical, and mathematical modeling to trapping records from the Hudson's Bay Company for hares and lynxes, as well as to extensive detailed field data, to suggest that phase dependencies in hunting success by lynx through the cycle resulted in nonlinear processes in the predator–prey interaction.

A difficult issue, however, is what inference can be drawn when, as is often the case, sparse and possibly non-independent time series are used for characterizing feedback, estimating parameters, and evaluating the performance of parameterized models. Does obtaining a high correlation between model output and observed changes in abundance (Kendall et al. 1999) indicate that the process most likely to drive the dynamics of the species has been identified? Is it possible to rank the plausibility of competing hypotheses on the basis of their relative ability to reproduce trajectories? Does the failure to achieve a high correspondence between predicted and observed population trajectories amount to a refutation of a hypothesis or the model? Everyone agrees that, when a precise quantitative match between observational data and model prediction is observed, the hypothesis is supported until a confounding variable that would explain this match is identified. However, such a quantitative approach is only rarely possible due to uncertainty in parameter estimation, model structure, and sampling error. Here, we take the view that, like any descriptive data, time series analysis is intrinsically unable to separate correlation from causation although, when appropriate data are available, it is a powerful tool for describing patterns, generating and refining hypotheses, and making predictions useful in a management context. However, a close correspondence between predictions and observations might be achieved for the wrong reason. Thus, where a quantitatively good fit can be obtained with time series data, the case for performing experiments is reinforced, not weakened. In the absence of experimental tests, inferring the causes of any natural phenomenon from time series analysis alone is an act of faith. Our skepticism about the kind of inference that can be drawn from TSA should not be confused with a negative attitude to the approach. Indeed, several of us have embraced the approach with the above objectives in mind (Steen et al. 1990, Stenseth et al. 1997, 1998, 1999, Lambin et al. 1998, Watson et al. 1998, 2000, Yoccoz et al. 1998).

Below, we elaborate in turn on three main limitations of time series analysis that restrict the strength of inferences that can be made on the basis of TSA, and on our unwillingness to stop short of performing experiments even when TSA-based investigation provides strong support for a hypothesis.

First, long time series are the product of a commitment to long-term monitoring of biological populations as well as the variables that affect them. Long time series are scarce and may lack measurements of things that we now think are part of the relevant ecological architecture. Thus, what we deduce about the structure of time series is based upon what we chose to start to measure many years ago. Hence, time series analysis is backward looking and likely to be confirmatory in nature instead of challenging theory. Most importantly, time series of sufficient length are highly biased in favor of large species and only rarely include macroparasites, even less microparasites. Thus, time series of the prevalence of bacteria and viruses are nearly nonexistent (but see Begon et al. 1999). Robust tests of, for example, the sign of partial derivatives requires long time series from populations with low exogenous variation or sampling noise. The larch budmoth example has by far the lowest exogenous variation of 14 insect time series, and the series is a very long one (Turchin and Taylor 1992). Whereas fitting mathematical models to Nicholson's laboratory blowfly populations time series uncovered the correct answer and illustrated the real potential of the approach (Kendall et al. 1999), the ratio of signal to noise in field populations is likely to be substantially lower. Thus, whereas the well-known classic data sets of population ecology such as Nicholson's blowflies, with both long time series and low exogenous noise, are of some interest, single and even multispecies time series data from most animal populations are lacking.

Second, the process(es) responsible (necessary and sufficient) for the cyclical nature of population dynamics may not be those accounting for most of the variance in population growth. Hence, relying on the capacity of a model parameterized from a time series to reproduce that time series cannot unambiguously identify the cause of cyclicity. Chapter 8 illustrates this issue well. One interpretation of the time series on autumnal moth is that all populations experience decadal peaks, but at most localities these are so low as to remain unnoticed. Thus, unless data are pooled over wide geographical areas, time series for this species will necessarily be dominated by the occasional large outbreaks. Ruohomäki et al. (2000) considered that climatic factors have the potential to turn low-amplitude oscillations into occasional highly visible outbreaks. Because one model of parasitism with parameters estimated from time series, and further fine-tuned, accounted for 55% of the variance in the same time series and generated cycles of appropriate periods, they favored parasitism as a likely explanation. But if two processes are involved in generating the cyclical dynamics and the occasional outbreaks of autumnal moth, using the capacity of single-factor models to reproduce the observed dynamics seems unproductive and likely to lead to incorrect conclusions.

Third, all time series consist of *estimates* of true abundance. They are often not error-free, not based on statistically robust estimation methods such as capture–recapture, nor a sample of any previously defined target population. Yoccoz et al. (2001) reviewed the consequences of these limitations for the nature of the inferences that can be drawn. Given these sources of bias, relying on quantitative agreement between model simulations and time series to rank and eliminate hypotheses is a risky exercise. The following four issues are of greatest concern.

(1) The longest time series data from vertebrates are derived from harvesting schemes and hence contaminated to various extents by *sampling error* and *bias*. Such biases at low density are common in wildlife harvest data (Royama 1992). For instance, the amplitude of snowshoe hare densities in intensive studies in the Yukon is approximately 14-fold (chapter 4), whereas fur return statistics suggest amplitudes in excess of 1000-fold. Counts of the number of red grouse hens in spring in northern England show amplitudes of 6-fold (Hudson et al. 1992), whereas shooting bag statistics suggest amplitudes in excess of 1000-fold (Hudson et al. 1998). Time series of the abundance of forest insects and their parasitoids are more straightforward to collect, although they also tend to be based on indices of abundance that have rarely been shown to be unbiased estimates of the true population size—hence, these suffer from similar, if less acute, problems as vertebrate harvest indices. For all species, the greatest source of discrepancy between real and estimated fluctuations is the difficulty of sampling species at low density (Hudson et al. 1999).

(2) Time series are typically *not spatially replicated*. Thus measures of uncertainty around any estimates of parameters derived from time series not only include sampling and process error, but also do not consider spatial variability, itself a further source of biases. For instance, Turchin's (chapter 7) "final verdict" that larch budmoth cycles are caused by feedback interactions with both parasitoids and plant quality is based on the analysis of a single, unreplicated time series of parasitoid abundance, and a single, unreplicated set of measurements of larch needle length taken in a different plot some

unspecified distance away from the former sampling area. Without any estimation of spatial variation in the process, reaching such a conclusion is as unwarranted as drawing statistical inference from a nonreplicated experiment. Instead, the analysis is extremely useful in formulating a strong hypothesis, and the testable prediction that neither removing parasitoids nor preventing changes in food abundance should stop the cycle.

(3) The precision with which relevant variables can be measured differs, precluding quantitative comparison of how well models of different processes reproduce time series. For instance, the relationship between an index of a relevant variable for which data are available (e.g., larch needle length) and the dynamically important variable (e.g., larch budmoth larval nutrition) may be less than perfect. Thus, even if the right model was specified and perfectly parameterized, time series models or models of hypotheses would necessarily account for a lower proportion of the variation in the rate of change of the proxy variable than would an equally appropriate model that includes a variable more directly related to the process modeled. In these circumstances, finding a weak correlation in a time series cannot be used to reject a causal hypothesis, nor can finding a weaker relationship for one variable than for another be used to conclude that the process in which it is involved is subsidiary (chapter 1).

(4) The type of dynamic pattern exhibited by a population may change over time or be nonstationary. When long runs of abundance estimates allow for their detection, long-term trends in dynamics are commonplace. They have been observed in snowshoe hares (Sinclair et al. 1993, Ranta et al. 1997), small mammals (Hanski and Henttonen 1996, chapter 3), and red grouse (chapter 6, Moss et al. 1996, Watson et al. 1998). Explanations range from near chaotic dynamics (Hanski et al. 1993), to variations in the strength of entraining exogenous cycles (potentially giving rise to *phase-forgetting* exogenous cycles such as with tree rings and snowshoe hares, Sinclair et al. 1993), changes in exogenous (precipitation) variables influencing dynamic variables (larval fecundity) (chapters 2 and 8), the inherent complexity of multispecies prey–predator assemblages (Hanski and Henttonen 1996), or following invasion by an alien predator (Oksanen and Henttonen 1996). Thus, true parameter values may drift over time, or the process responsible for the dynamics may change. If clear-cut changes in dynamics occur, and if time series data are available under each set of conditions, as is now the case with microtine rodents in Finnish Lapland (Henttonen 2000, chapter 3, and figure 9.1a), TSA may serve to test hypotheses by comparing the structure of time series of *Clethrionomys* voles in the absence of *Microtus* voles and of weasels, whose interaction is said to entrain the cycles of *Clethrionomys* voles.

Whereas statistical developments may help address some of the above issues by explicitly including the sampling process in the estimation of parameters of theoretical models fitted to data (e.g., Bjørnstad et al. 1998b), this is most often overlooked. While there is no doubt that substantial advances have been made in producing models that give good predictions of population trajectories, the limitations of time series analysis indicate the use of a pluralistic approach, which includes descriptive, statistical, theoretical, and experimental procedures.

9.5 Experimental Tests of Underlying Assumptions

The only known way to separate causation from spurious correlation is to resort to experiments. In the present context of population cycles, the most clear-cut experiment that would demonstrate the role of a trophic interaction in causing population cycles in given study populations would be to arrest these oscillations by removing or suitably interfering with some aspects of the interactions. Such an experiment would demonstrate that the presence of predators was *necessary* for population cycles to occur in the experimental setup. However, experimental tests of population-level predictions of hypotheses on population cycles, performed at a meaningful spatial scale (May 1999) and with acceptable levels of replication (Krebs et al. 1993, Hudson et al. 1998), require enormous effort and cost. Even so, if designed to test predictions that exclude other hypotheses, and if accompanied by accurate measurements of demographic parameters (chapter 4), population-level experiments remain the most promising avenue for separating correlation from causation.

Given their cost, it is more efficient if population-level experiments are preceded by strenuous attempts to test the plausibility of hypotheses by testing their logic through analysis of available time series, and modeling and testing their constituent assumptions through smaller scale experiments. Whereas hypotheses may have arisen from time series modeling, several of the hypotheses examined in this book had been formulated long ago, based on natural history observations (chapters 4 and 6). This approach is followed in chapters 3–6, although not all research programs traveled equally far along the road of attempted refutation.

Experiments attempting to refute the basic assumptions of a hypothesis are invariably easier to perform than attempts to modify population trajectories by manipulative experiments. It is therefore likely to be more efficient to do them first. That the empirical refutation of the assumptions of a hypothesis precludes the need to test its predictions has been little appreciated. The consequence of prematurely focusing on testing the population-level predictions of a hypothesis in the absence of support for its assumption is well illustrated by Krebs' (1978) genetic-based version of Chitty's polymorphism hypothesis. Over 20 years, much effort was expanded in testing the prediction of genetic changes in vole populations (Krebs 1970, 1979, Gaines and Krebs 1971) before any support for the assumptions of the heritability of life history and behavioral traits was available. Only 30 years after the hypothesis was formulated was the key experiment performed, and the assumption of a strong heritability of life history traits rejected (Boonstra and Boag 1987). A recent transplant experiment with field voles provided a further refutation of this and other hypotheses assuming that life history traits reflect the past environment experienced by voles, through the influence of maternal developmental effects or genetic inheritance, and not their present environment (Ergon et al. 2001).

A powerful way to strengthen the plausibility of a given mechanistic model is to provide experimental and statistical support for its constituent assumed relationships. In fact, McCallum (2000) considers that it is difficult to generalize from ad hoc models, which he defines as those not relying on experiments for characterizing functional responses and on statistical analysis of extensive data for parameter estimation. Whereas McCallum (2000) used the model of vole–weasel–generalist predator by Turchin and Hanski (1997) as his prime example of such an ad hoc model, Sundell et al. (2000) recently provided experimentally derived parameter estimates for the functional response of weasels that deviated only slightly from those assumed initially. A further five parameters in this seven-parameter model still require formal estimation. Given the spatial variation in the number and diversity of prey species available to predators, testing the assumption that functional responses are constant would also be valuable. We are not aware of any instance where the measurement of functional responses has been replicated outside laboratory conditions.

Testing whether parameter values assumed when optimizing the fit between simulated data and time series are compatible with empirical data is another way to test mechanistic hypotheses. For instance, Münster-Swendsen (chapter 2) reports that a complex model based on k-factors, and assuming that 75% of attacks by parasitoids of spruce needleminer result in pseudoparasitism, accounts for 78% of the variation in the per-capita rates of change of needleminers. Note that two-species logistic modeling captured the effect of parasitoids on the per-capita rate of change of the needleminer without any knowledge of how its birth rate was affected. On that basis, he concluded that "it is difficult to escape from the conclusion that the specific cause of population cycles is its interaction with parasitoids, including the effect of pseudoparasitism on the birth rate." In fact, he formulated a quantitative hypothesis based on a parameter value and an assumed relation of direct proportionality between the prevalence of pseudoparasitism and parasitoid abundance. This quantitative hypothesis should now be tested using ovary dissection in field populations, preferably not in the same population as that used to formulate the hypothesis, so as to provide evidence on the generality of the assumed relationship.

Models that predict population size invariably generate other, intermediate, predictions such as the size of constituent parameters. Testing such predictions with experiments or other data is yet another way to subject a hypothesis to proper empirical scrutiny. One such prediction of standard two-species predator-prey models is that predators should have a delayed density-dependent response to their prey because of the difference in growth rates or developmental delays. This intermediate prediction may not hold in real data if the predator is itself the prey of a higher level predator, or if the predator population is subjected to strong density dependence. For instance, current models of weasel–vole interactions assume that weasel survival is strictly density-independent (Turchin and Hanski 1997). Weasel territoriality resulting in density-dependent dispersal or predation on weasels by larger, generalist predators are two plausible mechanisms that would explain the absence of any delay in the numerical response of weasels (Graham and Lambin unpublished results). Because a delayed numerical response is required for predator–prey interactions to cause cycles, the failure to observe such a response in a cyclical population amounts to a refutation of the predator–prey hypothesis as a necessary process for cyclicity for a given system, without having to resort to a population-level experiment. The experiment reported in chapter 5 provides support for such a key "intermediate prediction" of the hypothesis that cycles in southern pine beetle (*Dendroctonus frontalis*) are caused by predator–prey interactions. Turchin et al. (1999) estimated the proportion of bark beetle larvae, protected from predators or exposed to natural predation, surviving to become emerging adults over the course of a whole 6-year cycle. Predator-imposed mortality was negligible during the increase phase, grew during the year of peak population, and reached a maximum during the period of population decline. Whereas the delayed pattern was as expected if predation by natural enemies played a driving role in driving *D. frontalis* oscillations, Reeve and Turchin (chapter 5) are well aware that further experiments would be required to rule out other mechanisms that can generate delayed density dependence. Nevertheless, the experiment was a useful step.

Strong exogenous influences may also weaken the underlying negative feedback that would otherwise arise in a trophic interaction. For example, Moss et al. (1993) found that most of the year-to-year variation in red grouse parasite burdens was explained by rainfall with only a weak influence of red grouse density, implying that processes other than parasitism were responsible for the grouse cycle in that study area. Thus, dissecting hypotheses on the cause of cyclicity into components and subjecting each in turn to empirical scrutiny has the potential to exclude false hypotheses without resorting to the more difficult and expensive large-scale experiments. Studies of lynx and coyote numerical and functional responses are exemplary in providing field data on components of predation on cyclic snowshoe hares (O'Donoghue et al. 1997, 1998). An investigation into the causes of cyclicity in a given system is, however, not complete until a hypothesis that survives preliminary tests is subjected to the ultimate test of its population-level predictions.

9.6 Experimental Test of Population-level Predictions

Despite calls for more experiments on cycles having been made repeatedly from many corners (Hanski and Korpimaki 1995, Korpimaki and Krebs 1996, May 1999, chapter 1), remarkably few have been attempted and two of the most significant are reported in this book (chapters 4 and 6, but see also Moss et al. 1996). Both experiments had as the main objective to transform periodic oscillations into nonperiodic erratic fluctuations, and tested the hypotheses that parasitism and predation were necessary and sufficient processes responsible for population cycles in red grouse and snowshoe hare, respectively. If these experiments were to be judged by their ability to put an end to controversies, they could be viewed as failures. The two experiments differ greatly in the level of replication achieved and in the nature and intensity of the response variable measured, yet each has been informative. Together, they highlight important pitfalls and suggest what experiments are likely to be most fruitful in the future.

The experiment by Hudson et al. (1998) is exceptional in being well replicated, conducted on a large spatial scale, based on a relationship demonstrated experimentally (Hudson et al. 1992), and testing predictions generated, although a posteriori, by a parameterized mechanistic model. The last feature is exceptional for field experiments with cyclic populations, although more readily achieved with laboratory populations (e.g., McCauley et al. 1999). Hudson et al. used anthelmintics to reduce worm burdens in cyclic red grouse managed for shooting. They drugged grouse in the springs of 1989 and 1993, each the first year of two separate declines. On two areas birds were drugged in 1989 and 1993, on two areas birds were drugged in 1989 only, and two areas served as controls (no drug). The response variable, grouse population size, was estimated indirectly as the number of grouse shot. No demographic data such as grouse survival or emigration, or worm burden before or after the experiment, were recorded. The number of birds shot declined 2- to 4-fold on the treated areas relative to peak time (figure 9.1b). No attempt was made to shoot grouse in low-density troughs in five out of the six untreated controls because counts of brood size in mid-July indicated to estate managers that too few grouse were present for shooting to be economical. The problem is that, despite having performed a large experiment, comparatively little energy was expended in accurately measuring the response variables. Had data on changes in worm burdens on control or experimental areas, or data on grouse survival, fecundity, or emigration been recorded, the experiment would have been more decisive (Lambin et al. 1999). As a general rule, and given that levels of replication for large-scale experiments will always be limited, collecting demographic and other ancillary data will always increase the value of experiments. Some measure of the magnitude of any experimentally induced difference between treatments and controls in the prevalence and burden of pathogens, or the abundance of predators, is also important if models are to be used to help interpret ambiguous outcomes of experiments. Although Hudson et al. point out that a decrease in cycle amplitude is the outcome predicted by a model of the parasitism hypotheses, this prediction is not unique to the parasitism hypothesis (see section 9.2). This ambiguity of outcome is likely to arise with any experiment designed to reduce the amplitude of the cycle rather than completely suppressing it. Thus, it is important to determine (with models) the magnitude of the treatment to be imposed so as to generate clear-cut predictions. The use of a parameterized model to interpret the outcome of the experiment raises another general issue. The model used to generate predictions (Hudson et al. 1998) did not include any stochasticity. It predicted that dampening cycles, similar to those observed, will occur following an approximately 20% reduction in worm prevalence. However, it is well established that, in model simulations, dampening cycles may be perpetuated by stochastic noise (e.g., chapters 1 and 2). Given that stochastic noise in parameter values is always present in natural populations, it is unclear whether real cycles should dampen or persist when experimentation with deterministic models predicts a dampening cycle.

The experiment reported by Boutin et al. (chapter 4) on the snowshoe hare cycle in the southwestern Yukon employed a set of manipulations of predator abundance, food supplies, nutrient additions to vegetation, and a combined predator manipulation with food addition. As such, it was a large-scale experiment done with no explicit mathematical model, but based on a factorial approach to experimentation. Predictions from previous studies by Keith and associates (Keith 1983) were stated before the experiments were carried out (Krebs et al. 1993), but a formal model of the system is only now available, after the experiments were completed (King and Schaffer 2001). A strength of the Kluane experiments was the detailed demographic data gathered on all the major vertebrate species in the ecosystem and, as such, it was in some sense the converse of the red grouse experiment. The implicit aim of the manipulations was to stop the 10-year cycle of snowshoe hares. This was not achieved, as reviewed by Boutin et al. in chapter 4, and the key arguments for the conclusion that predation is both necessary and sufficient to cause the hare cycle were based on changes in demographic parameters under the different experimental treatments. Density changes in fact were misleading in many of the treatments, because of local immigration and emigration, a problem even with relatively large-scale manipulations. The Kluane experiments did not conform to the view that a mathematical model is a necessary step before experimental manipulations are designed, and it is not clear what other manipulations would have been suggested by modeling exercises. Indeed, the alternative and more traditional view that motivated the Kluane experience is that we need first to do experiments on vague, verbal hypotheses and then proceed to model the system to search for missing assumptions to test at a later date.

9.7 Conclusions

Whereas it may seem obvious to many that the whole toolkit available to ecologists should be used to tackle the issue of population cycles, field experiments with cyclic populations are difficult because of the spatial and temporal scale over which they take place. Instead of believing that experiments on population cycles will necessarily produce indeterminate results, we favor striving to overcome the problems of implementation, together with the collection and use of time series to generate hypotheses, and model building to explore their plausibility. In fact, we are in complete agreement with Turchin

(1999) when he writes that "progress in ecology results from collecting and analyzing data using innovative techniques, developing insightful mathematical models, performing well-designed experiments, and, particularly, from interaction between the different approaches." We strenuously disagree with the view that experiments are inherently ambiguous and that support for a hypothesis may be so strong that experiments are unnecessary.

Whereas the reasoning underlying the experimental approach may be simplistic, it has the benefit of providing a basis for generating clear, testable predictions for experiments. If otherwise strongly supported hypotheses were consistently not supported by the failure to stop cycles in well-executed experiments, this would possibly lead to the rejection of the experimental paradigm and its replacement by a less simplistic alternative, such as a search for the feedback structure that dominates the observed dynamics advocated by Berryman (chapter 1). In our view, however, to date, too few experimental tests of population-level predictions of simple hypotheses on population cycles have been performed satisfactorily to justify abandoning the experimental paradigm.

Many chapters in this book end with the conclusion that trophic processes are *important* in generating the observed dynamics. Whereas such conclusions have the benefit of being uncontroversial—who would argue to the contrary?—ecology suffers from having no operational way of deciding if a factor is "important." The trophic hypotheses considered in this book have all been shown to be plausible, in many cases the most likely hypotheses for the cyclical nature of the observed dynamics. So far, in all cases, they remain just that: plausible hypotheses. We argued that using an approach based on the search for processes that are necessary for the cyclical nature of dynamics is more rigorous than searching for processes able to account for a high proportion of the variance in change in population size. By demonstrating causal links between processes, experiments offer a firmer basis than correlational evidence on which to build theories. This is because correlations usually have more than one possible explanation. Finally, a major virtue of experiments is that they can provide unexpected results, which can lead to the development of new hypotheses. The real world is much more surprising than models have envisioned, and forces us to consider ideas not previously imagined, indeed, as Mark Twain (1897) wrote, "Truth is stranger than fiction, but it is because Fiction is obliged to stick to possibilities. Truth isn't."

ACKNOWLEDGMENTS

We would like to thank Alan Berryman for putting up with our controversial views, also for numerous discussions and comments. We thank Peter Hudson and Heikki Henttonen for making their published data readily available to us. Karen Hodges provided useful comments on an earlier version, and comments by Judy Myers and Matt Ayres on chapters 1 and 2 provided useful insights. X.L. was supported by the Leverhulme Trust and NERC.

REFERENCES

Begon, M., S. M. Hazel, D. Baxby, K. Bown, R. Cavanagh, J. Chantrey, T. Jones, and M. Bennett. 1999. Transmission dynamics of a zoonotic pathogen within and between wildlife host species. *Proc. Roy. Soc. Lond., Ser. B.* 266: 1939–1945.

Berryman, A. A. 1982. Biological-control, thresholds, and pest outbreaks. *Environ. Entomol.* 11: 544–549.

Berryman, A. A. 1993. Food-web connectance and feedback dominance, or does everything really depend on everything else? *Oikos* 68: 183–185.

Bjørnstad, O. N., N. C. Stenseth, T. Saitoh, and O. C. Lingjaerde. 1998a. Mapping the regional transition to cyclicity in *Clethrionomys rufocanus*: spectral densities and functional data analysis. *Res. Popul. Ecol.* 40: 77–84.

Bjørnstad, O. N., M. Begon, N. C. Stenseth, W. Falck, S. M. Sait, and D. J. Thompson. 1998b. Population dynamics of the Indian meal moth: demographic stochasticity and delayed regulatory mechanisms. *J. Anim. Ecol.* 67: 110–126.

Boonstra, R. and P. Boag. 1987. A test of the Chitty hypothesis: inheritance of life-history traits in meadow voles *Microtus pennsylvannicus*. *Evolution* 41: 929–947.

Chitty, D. 1960. Population processes in the vole and their relevance to general theory. *Can. J. Zool.* 38: 99–113.

Ergon, T., X. Lambin, and N. C. Stenseth. 2001. Life-history traits of voles in a fluctuating population respond to the immediate environment. *Nature* 411: 1043–1045.

Fitzgerald, B. 1977. Weasel predation on a cyclic population of the montane vole (*Microtus montanus*) in California. *J. Anim. Ecol.* 46: 367–397.

Gaines, M. and C. J. Krebs. 1971. Genetic changes in fluctuating voles populations. *Evolution* 25: 702–723.

Hanski, I. and H. Henttonen. 1996. Predation on competing rodent species: A simple explanation of complex patterns. *J. Anim. Ecol.* 65: 220–232.

Hanski, I. and E. Korpimäki. 1995. Microtine rodent dynamics in northern Europe—parameterized models for the predator–prey interaction. *Ecology* 76: 840–850.

Hanski, I., L. Hansson, and H. Henttonen. 1991. Specialist predators, generalist predators, and the microtine rodent cycle. *J. Anim. Ecol.* 60: 353–367.

Hanski, I., P. Turchin, E. Korpimäki, and H. Henttonen. 1993. Population oscillations of boreal rodents—regulation by mustelid predators leads to chaos. *Nature* 364: 232–235.

Henttonen, H. 2000. Long-term dynamics of the bank vole *Clethrionomys glareolus* at Pallasjarvi, northern Finnish taiga. *Polish J. Ecol.* 48: 87–96.

Hudson, P. J. 1986. *The red grouse, the biology and management of a wild gamebird.* The Game Conservancy Trust, Fordingbridge, UK.

Hudson, P. J. 1992. *Grouse in space and time.* The Game Conservancy Trust, Fordingbridge, UK.

Hudson, P. J., D. Newborn, and A. P. Dobson. 1992. Regulation and stability of a free-living host–parasite system—*Trichostrongylus tenuis* in red grouse. 1. Monitoring and parasite reduction experiments. *J. Anim. Ecol.* 61: 477–486.

Hudson, P. J., A. P. Dobson, and D. Newborn. 1998. Prevention of population cycles by parasite removal. *Science* 282: 2256–2258.

Hudson, P. J., A. P. Dobson, and D. Newborn. 1999. Population cycles and parasitism: response. *Science* 286: 2425a.

Keith, L. 1983. Role of food in hare population cycles. *Oikos* 40: 385–395.

Kendall, B. E., C. J. Briggs, W. W. Murdoch, P. Turchin, S. P. Ellner, E. McCauley, R. M. Nisbet, and S. N. Wood. 1999. Why do populations cycle? A synthesis of statistical and mechanistic modeling approaches. *Ecology* 80: 1789–1805.

King, A. A. and W. M. Schaffer. 2001. The geometry of a population cycle: a mechanistic model of snowshoe hare demography. *Ecology* 82: 814–830.

Korpimäki, E. and C. J. Krebs. 1996. Predation and population cycles of small mammals—a reassessment of the predation hypothesis. *Bioscience* 46: 754–764.

Korpimäki, E. and K. Norrdahl. 1998. Experimental reduction of predators reverses the crash phase of small-rodent cycles. *Ecology* 79: 2448–2455.

Krebs, C. J. 1970. Genetic and behavioural studies on fluctuating vole populations. *Proc. Adv. Study Inst. Dynam. Numb. Popul.*: 243–256.

Krebs, C. 1978. A review of the Chitty hypothesis of population regulation. *Can. J. Zool.* 56: 2463–2480.

Krebs, C. 1979. Dispersal, spacing behaviour, and genetics in relation to population fluctuations in the vole *Microtus townsendii*. *Fortschr. Zool.* 25: 61–77.

Krebs, C. J. 1996. Population cycles revisited. *J. Mammal.* 77: 8–24.

Krebs, C. J. and K. DeLong. 1965. A *Microtus* population with supplemental food. *J. Mammal.* 46: 566–573.

Krebs, C. J., R. Boonstra, S. Boutin, M. Dale, S. Hannon, K. Martin, A. Sinclair, J. Smith and R. Turkington. 1993. What drives the snowshoe hare cycle in Canada's Yukon? In D. McCullough and R. Barret (Eds.) *Wildlife 2001: Populations*. Elsevier Science, London, pp. 886–896.

Lambin, X., D. A. Elston, S. J. Petty, and J. L. MacKinnon. 1998. Spatial asynchrony and periodic travelling waves in cyclic populations of field voles. *Proc. Roy. Soc. Lond., Ser. B.* 265: 1491–1496.

Lambin, X., C. J. Krebs, R. Moss, N. C. Stenseth, and N. G. Yoccoz. 1999. Population cycles and parasitism. *Science* 286: 2425a.

Lambin, X., S. J. Petty, and J. L. MacKinnon. 2000. Cyclic dynamics in field vole populations and generalist predation. *J. Anim. Ecol.* 69: 106–118.

Leslie, P. and R. Ranson. 1940. The mortality, fertility and rate of natural increase of the vole (*Microtus agrestis*), as observed in the laboratory. *J. Anim. Ecol.* 9: 469–477.

Litvin, K. Y. and N. G. Ovsyanikov. 1990. Relationship between the reproduction and numbers of snowy owls and arctic foxes and the numbers of true lemmings on the Wrangel Island. *Zool. Zh.* 69: 52–64.

MacLean, S., B. Fitzgerald, and F. Pitelka. 1974. Population cycles in arctic lemmings: winter reproduction and predation by weasels. *Arctic Alpine Res.* 6: 1–12.

May, R. 1999. Crash tests for real. *Nature* 398: 371.

McCallum, H. 2000. *Population parameters: estimation for ecological models.* Blackwell Scientific, Oxford.

McCauley, E., R. Nisbest, W. W. Murdoch, A. de Roos, and W. S. C. Gurney. 1999. Large-amplitude cycles of *Daphnia* and its algal prey in enriched environments. *Nature* 402: 653–656.

Moss, R. and A. Watson. 2001. Population cycles in birds of the grouse family (Tetraonidae). *Adv. Ecol. Syst.* 32: 53–111.

Moss, R., A. Watson, I. B. Trenholm, and R. Parr. 1993. Caecal threadworms *Trichostrongylus tenuis* in red grouse *Lagopus lagopus scoticus*—effects of weather and host density upon estimated worm burdens. *Parasitol.* 107: 199–209.

Moss, R., A. Watson, and R. Parr. 1996. Experimental prevention of a population cycle in red grouse. *Ecology* 77: 1512–1530.

O'Donoghue, M., S. Boutin, C. J. Krebs, and E. J. Hofer. 1997. Numerical responses of coyotes and lynx to the snowshoe hare cycle. *Oikos* 80: 150–162.

O'Donoghue, M., S. Boutin, C. J. Krebs, G. Zuleta, D. L. Murray, and E. J. Hofer. 1998. Functional responses of coyotes and lynx to the snowshoe hare cycle. *Ecology* 79: 1193–1208.

Oksanen, T. and H. Henttonen. 1996. Dynamics of voles and small mustelids in the taiga landscape of northern Fennoscandia in relation to habitat quality. *Ecography* 19: 432–443.

Oli, M. K. and F. S. Dobson. 2001. Population cycles in small mammals: the alpha-hypothesis. *J. Mammal.* 82: 573–581.

Pearson, O. 1966. The prey of carnivores during one cycle of mouse abundance. *J. Anim. Ecol.* 35: 217–233.

Ranta, E., V. Kaitala, and J. Lindstrom. 1997. Dynamics of Canadian lynx populations in space and time. *Ecography* 20: 454–460.

Royama, T. 1992. *Analytical population dynamics.* Chapman and Hall, London.

Ruohomäki, K., M. Tanhuanpää, M. P. Ayres, P. Kaitaniemi, T. Tammaru, and E. Haukioja. 2000. Causes of cyclicity of *Epirrita autumnata* (Lepidoptera, Geometridae): grandiose theory and tedious practice. *Popul. Ecol.* 42: 211–223.

Saunders, L., D. Tompkins, and P. J. Hudson. 1999. Investigating the dynamics of nematode transmission to the red grouse (*Lagopus lagopus scoticus*): studies on the recovery of *Trichostrongylus tenuis* larvae from vegetation *J. Helminthol.* 73: 171–175.

Sinclair, A., J. Gosline, G. Holdsworth, C. J. Krebs, S. Boutin, J. Smith, R. Boonstra, and M. Dale. 1993. Can the solar cycle and climate synchronize the snowshoe hare cycle in Canada? Evidence from tree rings and ice cores. *Am. Nat.* 141: 173–198.

Sinclair, A. R. E., C. J. Krebs, J. M. Fryxell, R. Turkington, S. Boutin, R. Boonstra, P. Seccombe-Hett, P. Lundberg, and L. Oksanen. 2000. Testing hypotheses of trophic level interactions: a boreal forest ecosystem. *Oikos* 89: 313–328.

Steen, H., N. G. Yoccoz, and R. A. Ims. 1990. Predators and small rodent cycles: an analysis of a 79-year time series of small rodent population fluctuations. *Oikos* 59: 115–120.

Stenseth, N. C., W. Falck, O. Bjørnstad, and C. J. Krebs. 1997. Population regulation in snowshoe hare and Canadian lynx: asymmetric food web configurations between hare and lynx. *Proc. Natl. Acad. Sci. USA* 94: 5147–5152.

Stenseth, N. C., W. Falck, K. S. Chan, O. N. Bjørnstad, M. O'Donoghue, H. Tong, R. Boonstra, S. Boutin, C. J. Krebs, and N. G. Yoccoz. 1998. From patterns to processes: phase and density dependencies in the Canadian lynx cycle. *Proc. Natl. Acad. Sci. USA* 95: 15,430–15,435.

Stenseth, N. C., K. S. Chan, H. Tong, R. Boonstra, S. Boutin, C. J. Krebs, E. Post, M. O'Donoghue, N. G. Yoccoz, M. C. Forchhammer, and J. W. Hurrell. 1999. Common dynamic structure of Canada lynx populations within three climatic regions. *Science* 285: 1071–1073.

Sundell, J., K. Norrdahl, E. Korpimäki, and I. Hanski. 2000. Functional response of the least weasel, *Mustela nivalis nivalis.* *Oikos* 90: 501–508.

Turchin, P. 1999. Population regulation: a synthetic view. *Oikos* 84: 153–159.

Turchin, P. and I. Hanski. 1997. An empirically based model for latitudinal gradient in vole population dynamics. *Am. Nat.* 149: 842–874.

Turchin, P. and A. D. Taylor. 1992. Complex dynamics in ecological time-series. *Ecology* 73: 289–305.

Turchin, P., A. D. Taylor, and J. D. Reeve. 1999. Dynamical role of predators in population cycles of a forest insect: an experimental test. *Science* 285: 1068–1071.

Twain, M. 1897. *Following the Equator.* American Publishing Co., New York.

Watson, A., R. Moss, and S. Rae. 1998. Population dynamics of Scottish rock ptarmigan cycles. *Ecology* 79: 1174–1192.

Watson, A., R. Moss, and P. Rothery. 2000. Weather and synchrony in 10-year population cycles of rock ptarmigan and red grouse in Scotland. *Ecology* 81: 2126–2136.

Williams, J. 1985. Statistical-analysis of fluctuations in red grouse bag data. *Oecologia* 65: 269–272.

Yoccoz, N. G., K. Nakata, N. C. Stenseth, and T. Saitoh. 1998. The demography of *Clethrionomys rufocanus*: from mathematical and statistical models to further field studies. *Res. Popul. Ecol.* 40: 107–121.

Yoccoz, N. G., J. D. Nichols, and T. Boulinier. 2001. Monitoring of biological diversity in space and time. *Trends Ecol. Evol.* 16: 446–453.

10

Do Trophic Interactions Cause Population Cycles?

Alan A. Berryman

10.1 Introduction

My motivation in editing this book has been to present as compelling and credible a story as possible. Although I am personally convinced of the soundness of our argument, that food web architecture plays a key role in the cyclic dynamics of many animal populations, I am not sure that others will be so convinced. In this final chapter, therefore, I exercise my prerogative as editor to have the last word, a final attempt to convince the skeptics and to answer the critics.

10.2 The Spruce Needleminer Cycle

Perhaps the most compelling case comes from the Mikael Münster-Swendsen monumental study of a needleminer infesting Danish spruce forests (chapter 2). Mikael is the only person I know of who has, almost single-handedly, and with considerable precision, measured all the variables suspected of affecting the dynamics of a particular population over an extended period of time (19 years) and in several different localities (seven isolated spruce stands). Others have longer time series from more places, but none has been so complete in terms of the number of variables measured. This exhaustive study enabled him to build a model of the complete needleminer life system, and use this model to home in on the factors responsible for the cyclical dynamics. However, the story would not have been complete without multivariate time series analysis, which led to the discovery of parasitoids as the cause of the key feedback process, density-related reduction in fecundity. The lesson from Münster-Swendsen's work is clear: If we want to understand population dynamics, we need long time series for all the variables likely to affect the dynamics of the subject population(s). In other words, we need to consistently monitor ecological systems over long periods of time and in many different locations. If there is a weakness in his study, it is the absence of the final definitive experiment. Such an experiment would be relatively easy and cheap to do (relative to those described in other chapters), because isolated spruce stands are common in Denmark and parasitoids emerge from the soil a week or two after the needleminer. Thus, parasitoids could easily be excluded by spraying the ground with an insecticide after needleminer emergence. One of the reasons why nobody seems prepared to do such an experiment is that Münster-Swendsen's data and analysis are so convincing that an experiment may seem to them redundant.

10.3 The Vole Cycle

Chapter 3 is a classical example of the application of hypothetical (mechanistic) modeling to an ecological problem and, as far as I know, it is the first time such an approach has been applied to the question of population cycles. Ilkka Hanski and his colleagues hypothesized that the vole cycle in northern Fennoscandia was caused by the interaction with specialist mustelid predators and stabilized in the south by generalist predators. The hypothesis seems to rest on sound empirical and theoretical foundations; that is, the facts that vole populations cycle in the north but not in the south, voles are preyed upon by effective specialist predators, and generalist predators are less abundant in the north and/or ineffective in winter because of deep snow. Also the theory that interactions with specialist predators can induce second-order (cyclical) dynamics, and that the switching and/or aggregation of generalist predators can stabilize prey populations and induce first-order (noncyclical) oscillations. Using classical functional-response equations, Hanski and his associates modeled the theoretical structure of the vole interaction with specialist and generalist predators and, using parameters obtained from independent data sources, demonstrated that the model could produce a variety of dynamics very similar to those observed in real life. While it is true that this study lacks statistical rigor, particularly in the way some parameter values had to be obtained in the absence of data (e.g., by educated guesses), the logical structure of the model and its basic results are difficult to ignore. Perhaps the most convincing argument is the ability of the model to account for all the observed patterns of vole fluctuations in Fennoscandia, something that other models have yet to do. However, we must remember that hypothetical models can only inform us if a particular hypothesis is plausible, or if it is more plausible than other competing hypotheses. What is missing from the vole story is convincing evidence that mustelids have a dominant impact on vole per-capita rates of change and that they have strong numerical responses to changes in vole density; that is, there is strong negative feedback between the two populations. This kind of evidence could be obtained by persistent monitoring of vole and predator populations and/or the execution of field experiments capable of separating the effects of different trophic relationships. What most vole ecologists seem to agree on nowadays is that the vole cycle is probably caused by some kind of trophic interaction. After half a century of research on intrapopulation (genetic or maternal) effects, this idea is now regarded as implausible because it cannot explain all the patterns of vole dynamics and, more importantly, has been falsified by recent experiments (e.g., Ergon et al. 2001a, b, Norrdahl and Korpimäki 2002).

10.4 The Snowshoe Hare Cycle

In chapter 4, Stan Boutin and his colleagues describe what must be the largest and most expensive long-term ecological experiment ever conducted. The problems of fencing and maintaining their 100 ha predator exclusion plot for 10 years under primitive conditions must have been a logistical nightmare. The experiments showed that: (1) fertilization and predator exclusion had little or no effect on mean hare densities or cycle amplitude, (2) food supplementation increased mean hare densities but had little effect on cycle amplitude, (3) predator exclusion plus food supplementation increased mean hare densities and dramatically reduced cycle amplitude. These results led the authors to conclude that the hare cycle must be the result of interactions between three trophic levels—the hare population, its food supply, and its predators—a conclusion that seems to be substantiated by the model of King and Schaffer (2001). (Note: It would have been more convincing, of course, if the model had been built, and had predicted the results, *before* the experiment was conducted.)

The snowshoe hare cycle has fascinated me ever since I noticed a strong third-order negative feedback effect in the Hudson Bay trapping records (see fig. 3.5 in Berryman 1981). In other words, a pure third-order model

$$R_t = 1.879 + 0.565 \ln N_{t-3} \qquad (10.1)$$

explained a remarkable 77% of the variation in hare per-capita rates of change. As Royama (1977) points out, however, this apparent third-order effect could be due to a mixed first- and second-order feedback structure. Fitting a second-order logarithmic model [equation (1.4)] to the same data yields the model

$$R_t = 1.062 + 0.571 \ln N_{t-1} - 0.889 \ln N_{t-2} \qquad (10.2)$$

which resolves 71% of the variation. This model is interesting because the first-order term is positive, indicating a positive effect of hare densities on its own rate of change. As far as I know, none of the current hypotheses or models considers such an effect.

The critical question, of course, is which model is correct: Some possibilities are (1) a pure third-order structure [$B \to C \to A \to B$ as implied by model (10.1); here A is food, B hares, and C predators], (2) first-order cooperation among hares plus second-order interaction with food [$B \to B$ plus $B \to A \to C \to B$ as implied by model (10.2)], (3) first order cooperation among hares plus second-order interaction with predators [$B \to B$ plus $B \to C \to B$ as could also be implied by model (10.2)], (4) first-order food limitation plus second-order interaction with food [$A \to A$ plus $B \to A \to B$], (5) first-order food limitation plus second-order interaction with predators [$A \to A$ plus $B \to C \to B$], or (6) a structure involving two second-order effects [$B \to A \to B$ plus $B \to C \to B$ as in the model of King and Schaffer (2001)]. Thus, although the experiments described by Boutin and his colleagues are consistent with the hypothesis that the hare cycle is a result of trophic interactions, the details of the trophic structure are still not apparent. Additional modeling of alternative hypothetical structures and, possibly, more focused experimentation will apparently be required before the details of the hare cycle are finally understood.

10.5 The Southern Pine Beetle Cycle

Chapter 5 illustrates how models can be used to design field experiments. Reeve and Turchin used a hypothetical model of the interaction between natural enemies and the southern pine beetle (SPB) to predict beetle mortality over an outbreak cycle in caged and exposed sections of loblolly pine stems, and then tested this prediction in a 5-year experiment. The results of the experiment are consistent with the hypothesis that natural enemies (mainly a predaceous clerid beetle) exerted a delayed negative feedback on the dynamics of the bark beetle and, therefore, could be involved in the cyclic oscillations.

This analysis and its results have been particularly intriguing to me since I have spent most of my life studying the population dynamics of bark beetles. The conventional view is that outbreaks of tree-killing (aggressive) bark beetles are triggered by increases in the abundance of susceptible hosts (e.g., trees weakened by storms, droughts, old age, or competition), and driven by the momentum of the beetle population (i.e., positive feedback due to the ability of large beetle populations to overcome the defenses of healthy trees by coordinated "mass attack" in response to pheromones) (see Berryman 1982 for a review). In contrast, Reeve and Turchin's analysis suggests that host resistance may not play such an important role in SPB dynamics.

The problem with the SPB time series it that they do not exhibit the rigid periodicity or synchrony characteristics of the other cyclic populations discussed in this book (figure 10.1). While some outbreaks seem to be synchronous across most southern states (e.g., outbreaks in 1972–75, 1979–81, 1984–87, and 1995), others are more restricted (outbreaks in Virginia in 1982 and 1993, Georgia in 1988, and Alabama in 1991–92). In addition, the period of

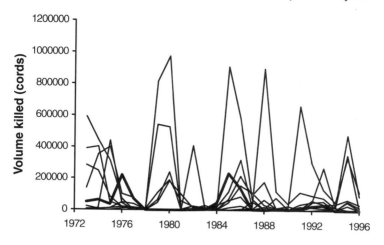

Figure 10.1 Oscillations in timber losses to the southern pine beetle by state in the southern United States. The thick line is the data from Texas used by Reeve and Turchin in chapter 5 (data compiled from Price et al. 1992).

the cycles is quite variable, ranging from 2 to 9 years and averaging 4.4 years. This is not the typical pattern of the classical population cycle (e.g., figure 2.2). However, analysis of the pheromone-trap time series (figure 5.4) using a two-species logistic model [equation (1.8)] indicates that there is fairly strong negative feedback between SPB and clerid populations; that is, the R-functions

$$R_b = 5.584 - 0.00402 B_{t-1} - 0.5879 C_{t-1}/B_{t-1},$$
$$R_c = 1.757 - 0.00046 C_{t-1} - 0.1302 C_{t-1}/B_{t-1},$$
(10.3)

with B and C representing pine beetles and clerids, resolved 68% and 71% of the variation, respectively (trap data from figure 5.4 was multiplied by 1000 before analysis). Interestingly, this model is very unstable, with the bark beetle population invariably being driven to extinction in a few years. This may not be altogether unreasonable, since no SPB have been caught in Texas in 2000 or 2001. Taken together with the evidence presented in chapter 5, it seems clear that clerid predators have a strong impact on SPB dynamics, perhaps driving them, periodically, to local extinction. This conclusion is bolstered by my experience as a graduate student studying the predators of the western pine beetle (WPB), a close relative of SPB. During this study I observed clerid populations responding numerically to WPB density and inflicting extremely high mortality on bark beetle adults and immature stages (Berryman 1970). Clerid larvae often reached very high densities (up to 40 per square foot), capable of completely eliminating a WPB brood under the bark, an impact that is rarely observed in the other bark beetles I have studied (e.g., fir engraver, mountain pine beetle, and European spruce beetle). This may be partly due to the relative attractiveness of the bark beetle pheromones to their respective clerid predators. For example, more clerids than bark beetles are usually caught in traps baited with SPB-aggregating pheromones (figure 5.4) and *Enoclerus lecontei* adults are observed in large numbers hunting on trees attacked very recently by WPB. However, clerids do not seem to respond so strongly to the pheromones of other bark beetles (Bakke 1982, Billings 1985). This suggests that certain clerids are particularly adapted (specialized) to search out trees attacked by WPB and SPB. Despite the significant impact of clerid predators on SPB dynamics, however, the irregularity and asynchrony of the cycles, plus the knowledge that conifers possess very effective defenses against bark beetles (Berryman 1972), leads me to suspect that host resistance, and perhaps other factors, also play a part in SPB dynamics. One possibility is that local outbreaks are triggered by reduced host resistance and terminated by clerid predation. Whatever the exact causes of SPB (and other bark beetle) population fluctuations, it is clear that trophic interactions of one sort or another play a critical role.

10.6 The Grouse Cycle

In chapter 6, Peter Hudson and his colleagues apply a spectrum of scientific approaches to reach the conclusion that red grouse population cycles in the United Kingdom are likely to be caused by the interaction with nematode parasites. Time series analysis indicated a dominant second-order effect. Models describing the interaction between grouse and nematode parasite populations predicted cycles of the appropriate period and amplitude. They then used the model to predict the effects of different antibiotic treatments aimed at reducing the level of parasitism. Finally, they tested these predictions by replicated field experiments. As the authors acknowledge, their interpretations have been severely criticized because the treatments failed to completely suppress the cycle. However, as Hudson et al. argue, the prediction of the model for the level of treatment actually applied was to suppress not eliminate the cycle. Although it is apparent that this argument is not going to disappear in the foreseeable future, Hudson et al.'s studies at least fail to *falsify* the parasite hypothesis. Thus, nematode parasitism must be taken seriously as a plausible explanation (perhaps the most plausible) for the grouse cycle. It is equally clear that these results do not disprove alternative hypotheses (see chapter 9). From my point of view, the one thing missing from the red grouse study is a structural analysis of time series data. It would be interesting, I think, to see how much of the variability in red grouse population change is associated with the abundance of the strongyle worm, and vice versa. In other words, how strong is the feedback between these two populations.

10.7 The Larch Budmoth Cycle

In chapter 7, Peter Turchin and his associates use a combination of time series analysis and hypothetical (mechanistic) modeling to evaluate two plausible explanations for the spectacular cycles of larch budmoth populations in the Swiss Alps. They first build a model of the interaction between the budmoth and its food supply using experimental data describing the effect of larch foliage taken from trees subjected to different degrees of past defoliation. The predictions of this model are qualitatively similar to the data, particularly in the period and amplitude of the oscillations. However, multiple time series analysis did not support the hypothesis of strong negative feedback between the budmoth population and foliage quality; that is, although there was a strong effect of budmoth density on needle lengths in the following year, the effect of needle length on budworm rates of change was very small (I obtained a similar result in chapter 1). This, together with the fact that needle lengths did not decline in the last budmoth outbreak, led them to conclude that budmoth–food interaction was not likely to drive the cycle. They then fit a Nicholson–Bailey-type model to the budmoth and parasitoid time series and found that it resolved a very high percentage of the variation (as I also found in chapter 1 with a logistic-type model). Even though this model produced less convincing dynamic behavior (the cycle being too short), its high resolution led the authors to conclude that parasites were mainly responsible for the cyclic oscillations. Turchin et al. then incorporated the two models into a three-trophic-level structure and found that the combined effects of parasitoids and food quality provided a better description of the observed cycles than either model alone. This led them to conclude that a three-trophic-level hypothesis was necessary to explain the details of the budmoth cycle.

Although the analyses presented in chapters 1 and 5 are both consistent with the hypothesis that larch budmoth cycles are the result of trophic interactions, their predictions concerning the forces that destabilize budmoth dynamics are quite different. My analysis predicts that interactions with parasitoids destabilize the dynamics and, therefore, are the main cause of the cycles; that is, if one removes parasitoids from the system, the cycles will be suppressed or eliminated. However, Turchin et al.'s model predicts that parasitoids actually stabilize the dynamics in the sense that, if parasitoids are removed from the system, the cycles will increase in amplitude and period. A critical experiment is needed to separate these hypotheses.

10.8 The Autumnal Moth Cycle

In chapter 8, Miia Tanhuanpää and her colleagues employ a hypothetical model to show that it is possible for interactions with parasitoids to cause population cycles of the autumnal moth in northern Fennoscandia. This is the third case in the book where population cycles of a forest defoliator seem to result from the interaction with insect parasitoids, and there may be good reasons why this should be so (see Berryman 1996). There are, however, some subtle differences in the behavior of these three populations. In the first case, the population dynamics of the spruce needleminer appear to be determined almost entirely by interactions with insect parasitoids (chapter 2). Unlike the larch budmoth and autumnal moth, needleminer populations rarely defoliate spruce stands in Denmark, so feedback to the host plant population is probably weak or nonexistent. In addition, the average period of the needleminer cycle (6 years) is considerably less than that of the others (9 years). Not surprisingly, most parasitoid–host models with reasonable parameter values (like those developed in chapters 7 and 8) predict cycles of around 6 years, while models containing feedback to predator and food populations predict longer cycles (chapter 7). This observation may lead to an important generalization: a cycle of period <8 implies a two-trophic-level structure while a cycle of period >8 implies a three-trophic-level structure (see also chapter 4). This generalization is supported by the autumnal moth time series, which may show both patterns (figure 8.1). Notice that, during the first outbreak, the moth population declined by 99% from 1955 to 1956 in the absence of parasitoids. The huge caterpillar population in 1955 (1.6 larvae per shoot) completely defoliated the trees before development was complete, and numerous larvae starved to death (Bylund 1995). Likewise, in the more modest second outbreak, the moth population also started its decline in 1965 without any parasitism. Even in the last outbreak, which had the lowest peak density of all, the population decline started in 1993 with less than 2% parasitism. In all these cases, the outbreak must have been stopped, and the decline initiated by interactions with food or some other unknown factor. The only outbreak where parasitoids may have played a significant role in terminating growth was the third (1987), when parasitism reached 50% at the peak and 96% the following year.

These observations are further supported by multiple time series analysis. Fitting a two-species logistic model [equation (1.8)] to the first series (1955–67) only explained 46% of the variance in autumnal moth rates of change, and 64% in parasitoid rates of change. Moreover, in both cases it was the first-order term (the self-regulation effect) that accounted for almost all of the variability (38% and 64%, respectively). However, when the same model was fitted to the second series (1984–94), the parasitoid/host ratio resolved 80% of the variation in autumnal moth R-values, and 89% in the parasitoid R-values. Moreover, neither coefficient of determination was improved by adding the first-order terms. In other words, the feedback between parasitoid and host populations appeared to be very strong during the second series and weak or nonexistent during the first. The fact that the period of the cycle was reduced from 10 years in the first series to 6 years in the second supports the generalization concerning the structure of cyclical populations.

10.9 Philosophical Issues

Most of the authors of this book subscribe to the view that population dynamics in general, and population cycles in particular (being only one aspect of the general subject), are often (perhaps usually) the consequence of ecological structures or food web architecture. Many of us belong to the *analytical* school of population dynamics, in that we tend to analyze and model data in our attempts to understand the causes of population fluctuations. This approach is particularly prevalent amongst applied ecologists, especially entomologists and fisheries biologists, and has a strong tradition in statistics and mathematics.

When I began putting this book together, and at the urging of our editor, I decided to try and add perspective by inviting somebody from a different school of thought to critically appraise the evidence we had compiled. I eventually asked Xavier Lambin who, I knew, had been critical of some of the examples in this book and is a strong advocate of the *experimental approach*. He recruited Charley Krebs, Robert Moss, and Nigel Yoccoz to help him with this task. That these authors, some of whom have been proponents of intrapopulation theories of population cycles, have chosen to accept, without serious protest, our arguments and evidence for the role of interpopulation (trophic) mechanisms, should be taken as a positive sign. However, they have raised some important philosophical issues about the roles of analysis and experimentation in the scientific method.

Being proponents of the experimental approach, the authors of chapter 9 believe that experiments are necessary to separate correlation from causation. They adhere to the Popperian view that experimental falsification is the only way to finally answer any specific scientific question. However, in their appraisal of the experiments described in this book, they are forced to admit that they do not always give clear-cut answers, for there may be more than one way to interpret experimental results. Meanwhile, the analytical approach has at least contributed to our general understanding of the necessary conditions for population cycles (see chapter 1). While it cannot "prove causation" (can anything?), analysis can certainly provide useful information on the relative effects of different ecological structures known a priori to fulfill the necessary conditions for population cycles. In other words, it seems perfectly reasonable to employ analytical methods to weigh the plausibility of different structural hypotheses. I think that most of the authors of this book would agree, however, that the final test should be an experiment designed to falsify these hypotheses, although this is not always possible because of time, logistical, and/or financial constraints.

Lambin et al. argue that experiments should be designed to search for processes that are *necessary* to cause the observed population cycles, an argument originally made by Dennis Chitty, another champion of the experimental approach (Chitty 1960). There is a problem, however, with the notion of "necessary and sufficient" conditions when applied to the causes of population change. First, everybody seems to agree that delayed negative feedback is the *general* necessary condition for an *endogenous* population cycle. The problem is that this general condition can be met by many specific structures; for example, interactions with food, predators, parasites, and pathogens, as well as certain intrapopulation processes like maternal effects. Although one of these processes may dominate the dynamics at any time and/or place, if it is removed by an experiment another may take over, so that the system still cycles (the hierarchy of control discussed by Berryman et al. 1987). From this standpoint, none of the potential delayed feedback structures is really necessary, because they all can all induce cycles, yet all are sufficient, for the same reason. In fact, the authors of chapter 9 seem to agree when they state, in reference to the rodent cycle, that there may be "many sufficient but no ... necessary conditions." It does not seem reasonable to design a research program around a search for necessary specific conditions that may not exist. A more effective strategy may be to search for the feedback structure that dominates the observed dynamics (chapter 1, Berryman 2001).

Lambin et al. regard the conclusions drawn by many of the authors of this book to be no more than plausible hypotheses. But, if we adopt the Popperian philosophy, that hypotheses can only be falsified, not proven, then all hypotheses, even those that survive experimental testing, are only plausible at best. In addition, the total body of the evidence presented in this book certainly conforms to the *theory* that endogenous population cycles can be caused by food web architecture, a theory that goes back at least as far as the predator–prey models of Lotka and Volterra. Theories based on pure logic, however, are unconvincing to some. To gain credence, the theory needs the support of empirical data. To this effect, the examples discussed in this book can be viewed as empirical support for the general theory of food web dynamics and its role in population cycles. This theory predicts that strong feedback between consumers and their resources can generate cyclic dynamics in both populations. However, it is not obvious how such a prediction can be falsified. The examples in this book at least provide evidence that cyclic populations are often characterized by strong feedback with their food and/or consumers.

ACKNOWLEDGMENTS

I would like to express my deep appreciation to the contributors to this book for their patience, and the scholarly manner in which they have conducted their disagreements with the editor. I would also like to express my sincere thanks to the many others who have selflessly contributed through the review of chapters and discussion of ideas. It has been a pleasure to work with you.

REFERENCES

Bakke, A. 1982. Mass trapping of the spruce bark beetle *Ips typographus* in Norway as part of an integrated control program. In A. F. Kydonieus and M. Beroza (Eds.) *Insect suppression with controlled release pheromone systems, II.* CRC Press, Boca Raton, Fla., pp. 17–25

Berryman, A. A. 1970. Evaluation of insect predators of the western pine beetle. In R. W. Stark and D. L. Dahlsten (Eds.) *Studies on the population dynamics of the western pine beetle,* Dendroctonus brevicomis *LeConte (Coleoptera: Scolytidae).* University of California, Berkeley, Division of Agricultural Sciences, pp. 102–112.

Berryman, A. A. 1972. Resistance of conifers to invasion by bark beetle–fungus associations. *Bioscience* 22: 598–602.

Berryman, A. A. 1981. *Population systems: a general introduction.* Plenum Press, New York.

Berryman, A. A. 1982. Population dynamics of bark beetles. In J. B. Mitton and K. B. Sturgeon (Eds.) *Bark beetles in North American conifers—a system for the study of evolutionary ecology.* University of Texas Press, Austin, Tex., pp. 264–314.

Berryman, A. A. 1996. What causes population cycles of forest Lepidoptera? *Trends Ecol. Evol.* 11: 28–32.

Berryman, A. A. 2001. Functional web analysis: detecting the structure of population dynamics from multi-species time series. *Basic Appl. Ecol.* 2: 311–321.

Berryman, A. A., N. C. Stenseth, and A. S. Isaev. 1987. Natural regulation of herbivorous forest insect populations. *Oecologia* 71: 174–184.

Billings, R. F. 1985. Southern pine bark beetles and associated insects: effects of rapidly-released host volatiles on response to aggregation pheromones. *Z. Angew. Entomol.* 99: 483–491.

Bylund, H. 1995. *Long-term interactions between the autumnal moth and mountain birch: the roles of resources, competitors, natural enemies and weather.* Ph.D. thesis, Swedish University of Agricultural Sciences, Uppsala, Sweden.

Chitty, D. 1960. Population processes in the vole and their relevance to general theory. *Can. J. Zool.* 38: 99–113.

Ergon, T., J. L. MacKinnon, N. C. Stenseth, R. Boonstra, and X. Lambin. 2001a. Mechanisms for delayed density-dependent reproductive traits in field voles, *Microtus agrestis*: the importance of inherited environmental effects. *Oikos* 95: 185–197.

Ergon, T., X. Lambin, and N. C. Stenseth. 2001b. Life-history traits of voles in a fluctuating population respond directly to environmental change. *Nature* 411: 1043–1045.

King, A. A. and W. M. Schaffer. 2001. The demography of a population cycle: a mechanistic model of snowshoe hare demography. *Ecology* 82: 814–830.

Norrdahl, K. and E. Korpimäki. 2002. Changes in individual quality during a 3-year population cycle of voles. *Oecologia* 130: 239–249.

Price, T. S., C. Doggett, J. M. Pye, and T. P. Holmes. 1992. *A history of southern pine beetle outbreaks in the southeastern United States.* Georgia Forestry Commission, Macon, Ga.

Royama, T. 1977. Population persistence and density dependence. *Ecol. Monogr.* 47: 1–35.

Index

Page numbers in *italic* refer to figures and tables.

acquired resistance, 114
adaptive management, 23
ants, 149
Apanteles tedellae, 30
Athous subfuscus, 30
autocorrelation, 11–12, 38, *54*, 111
  partial, 12, 111–112, *113*
autumnal moth, 142–154, 157, 162, 165, 183–184
avian predators, 53–55, 150

badgers, 53
bark beetles, 92–108, 180–182
behavioral effects, 110, 118, 127
Betula pubescens (*see* mountain birch)

Calluna vulgarus (*see* heather)
carrion crow, 122
cats, feral, 53
Cecidomyidae, 30, *31*
cerambycid, 102, *103*
chaos, 51, 166
Chitty effect, 50, 167 (*see also* genetic effects)
Circus cyaneus (*see* harrier, hen)

cleptoparasitoid, 30
clerid, 93–105, 180–182
Clethrionomys
  glariolus, *45*, 46, 157, *158*
  rufocanus, *45*, 46, 61
  rutilus, *45*, 46
climatic change, 61
climatic effects, 3, 56, 121, 143, 165
competition, 5, 41, 51, 102–104, 110
correlation, 14
  between time series, 14–18, 139, 160, 163
  spurious, 14, 164, 167
Corvus corone (*see* carrion crow)
coyote, 80–83, 84–85, 169
cross-correlation (*see* correlation, between time series)
cycle (*see* population cycle)

Dendroctonus
  brevicomis (*see* western pine beetle)
  frontalis (*see* southern pine beetle)
  ponderosae (*see* mountain pine beetle)
density dependence (*see* negative feedback)

diagnosis, 11, 20, 22, 38
Douglas-fir tussock moth, 8
driving variable, 6–8

eagle, bald, 80
Enoclerus lecontei, 104, 182
environmental variability, 8, 20, 61
Epinotia tedella (*see* spruce needleminer)
Epirrata autumnata (*see* autumnal moth)
equilibrium
  destabilization of, 110–111, 115, 126–127, 183
  stability of, 8, 20, 36, 37, 51, 53, 122, 149–150, 181
eradication, 116, *117*, 118
European spruce beetle, 182
experimental approach, 23, 25, 69–91, 93, 102–104, 116–121, 155–156, 160–172, 178, 182

fecundity effects, 31, 33, 35, 38–39, *40*, 41, 42, 111, 114, *115*, 134, 147, 149, 178
fence effects, 85
finite rate of increase, 134
fir engraver beetle, 8, 105, 182
food effects, *15*, 16–17, 20, 56, 69–88, 130–131, 131, 132, 133–137, 138–140, 149, 165–166
food webs, 4, 10–11
fox, red, 53, 122, 124
frontalin, 97
functional response, 22, 51, 84–85, 100, 122–123, 168, 169
fungus
  bluestain, 102–104
  disease, 30, *31*, 33, *34*, 35, 37

genetic effects, 4, 9, 24, 50, 58, 143–144, 149, 179
goshawk, 80–83
group selection, 110
gyrfalcon, 127

harrier, hen, 122–123, *124*, 126
harvesting effects, 125–126, 165
hawk, Harlan's, 80, 83
heather, 114, 121

host resistance, 114 (*see also* plant resistance)
hyperparasitoid, 30, 35

immigration, 100, 104
induced resistance, 4
Ips
  grandicollis, 95, *96*
  typographus, 105
Ixodes ricinus, 121

Keith hypothesis, 69, 85
key factors, 30, 33, 38, 39, 41
kin selection, 4, 120

larch budmoth, 6, 11, *12*, 14, *15*, 16–18, 130–141, 164, 165–166, 183, 184
lemming, Norwegian, 44, *45*, 46, 49–50, 56, 162
Lemmus lemmus (*see* lemming)
lichens, 143, 149
life tables, 22, 30–31, 33, 38–39
limiting factors, 160
loblolly pine, 92
louping ill virus, 121, 126
lynx, 69, 80–83, 84–85, 87, 163, 169

marten, 80
maternal effects, 4, 9, 24, 45, 57–58, 131–132, 143–144, 167, 179
Mattesia, 30, 33
metapopulation, 104
Microtus agrestis, *45*, 50, 61, 157
Microtus oeconomus, 50
mimicry, 10
models, 18–23, 162
  Anderson–May, 110, 114–118
  Beddington, 99, 137
  falsification of, 162
  Gompertz, 19
  hypothetical, 22, 24, 51–52, 166, 178, 180, 183
  logistic, 19–21, 38, 41, 51, 53, 98, 100, 104, 168, 181, 183, 184
  Lotka–Volterra, 19–21, 98
  mechanistic, 22, 168, 170, 178, 183
  Nicholson–Bailey, 100, 104, 137, 144, 183
  pseudointerference, 35

Ricker, 136, 144 (*see also* logistic model)
Rosenzweig–MacArthur, 53
systems, 22–23, 35–38, 177
Thompson, 100
time series, 18–22, 146, 160, 166
Moran effect 37, 55 (*see also* population cycle, synchrony of)
mountain birch, 142
mountain pine beetle, 105, 182
Mustela erminea, 50–54
Mustela nivalis, 50–54
mustelids, 45, 50–55, 80, 162–163, 166, 168–169, 178
mycosis (*see* fungus disease)

natural selection, 4
necessary conditions, 8, 160–161, 164, 167, 169, 170, 171, 172, 183, 185–186
negative feedback, 6, 8–11, 17, 99, 104, 137, 178, 182
  dominance, 11, 160, 186
  hierarchies, 23
  order of, 8, 10, 12–13, 15, 38, 104, 112, 130, 131, 133, 146, 178, 179–180, 182
  time delays in, 6, 8–10, 14, 19, 24, *34*, 85, 93, 95, 97, 102, 104, 114, 115, 126, 133, 144, 148, 150, 159–161, 168–169, 180
nematode, 109–129, 182
Norway spruce, 29
numerical response, 50, 51, 53, 69, 84–85, 88, 97, 105, 122, 126, 157, 161, 169, 178
nutrient cycling, 4

Operophtera brumata (*see* winter moth)
order of dynamics (*see* negative feedback)
owls, 53
  great horned, 80, 81–83
  hawk, 80
  Tengmalm's, *48*

Paecilomyces farinosus, 30
paradox of enrichment, 116

parasitoid effects, 14–18, 29–43, 101, 133, 137–140, 144–146, 150, 165, 178, 183–184
partial rate correlation function, 12–13, 38
partridge, rock, 126
pathogen–parasite effects, 3, 4, 56–57, 109–129, 114, *115*, 132, 162, 169, 170, 182
per-capita rate of change, 12, 14, 19–20, 114, 134, 144, 160, 178
pheromone, 92, 93, 97, 102, 180, 181, 182
physiology, 88
Picea abies (*see* Norway spruce)
Pimplopterus dubius, 30
Pinus echinata (*see* shortleaf pine)
Pinus taeda (*see* loblolly pine)
plant effects, 3, 4, 41, 45, 56, 57, 69–88, 101, 105, 133–137, 138–140, 146–148, 165, 183, 184
plant resistance, 4, 105, 146–148, 150, 180, 182
polydnavirus, 39
polymorphism, 132
Popperian, 161, 185, 186
population cycle
  amplitude of, 20, *32*, *33*, 46, 49, 52, 53, 55, 60, 61, 62, 71, 73, *76*, 86, 116, 118, *119*, 126, 130, 138, 139, 142, 145, 157, 159, 165, 170, 179, 182, 183
  causes of, 5–14, 20, 24
  dampening of, 115, 125, 126, 171
  definition of, 5, 24
  period of, 20, *32*, *33*, 41, 46, 52, 53, 112, 114, 115, 126, 130, 138, 139, 142, 145, 157, 159, 180, 182, 183, 184
  suppression of, 116–117, 183
  synchrony of, 31, *32*, 36, 49, 50, 55, 62, 143, 162, 180, 182
population regulation, 5, 9, 12, 150
positive feedback, 149, 180
predator effects, 3, 4, 5, 24, 45, 50–55, 69–88, 93, 95, *96*, 102, *103*, 148, 150, 157, 161, 162–163, 169, 170
predator pit, 123
predator/prey ratio, 14–15, 19–20
predator switching, 53, 178

prediction, 41–42, 116, 117, 125, 164, 166, 167, 168, 170, 180, 182, 183
protozoa, 30, *31*, 33, 35
pseudoparasitism, *31*, 39–40, 41, 42
ptarmigan
  rock, 127
  willow 126

R-function, 14–21, 93, 163, 181
random variation, 36, 42
raptors, 53, 81
ratio-dependence, 14, 15, 20, 51, 98, 100, 104
red grouse, 24, 109–129, 157–159, 162, 165, 166, 169, 170, 182
  aggregation of, 111, 120
  fecundity of, 111, 114, 117
  parasites of, 109–111
  survival of, 111, 114, *115*, 121–122
reindeer, 126
response surface methodology, 19, 93, 136
rodents 44–68, 157 (*see also* small mammals)

sampling, 30–31, 165
Scolytus ventralis (*see* fir engraver beetle)
sheep tick, 121
shortleaf pine, 92
simulation, *21*, 35–36, *37*, *52*, *54*, *101*, *117*, *125*, *135*, 136, 138, *139*, 145–146, *147*, 150, 168
small mammals, 150, 162, 166
snowshoe hare, 69–91, 163, 165, 166, 169, 170, 171, 179–180
  natality estimates, 71, 74–75, 77, 78
  survival estimates, 71, 78, *79*, 80, *81*, 82, *83*, *84*
Soay sheep, 126
social behavior hypothesis, 111, 118 (*see also* behavioral effects)
southern pine beetle, 24, 92–108, 169, 180–182
spacing behavior (*see* behavioral effects)
spatial scale, 25, 161, 167, 170, 171
spatial variability, 148–149, 157, 161–162, 165–166, 168
spectral analysis, 112, *113*
Sphagnum moss, 121

spruce needleminer, 5, 23, 29–43, 168, 177–178, 184
  fecundity, 31, 33, 35, 38–39, *40*, 41, 42, 178
squirrel
  ground, 84
  red, 83–84
  strong interactions, 11
sufficient condition, 8, 160–161, 164, 170, 171, 185–186
sunspots, 3, 56, 143
stoats (*see* mustelids)
sycamore aphid, 6, 11, *12*

Thanasimus dubius, 63–103
time series
  analysis of, 11–18, 25, 38–39, 41, 88–89, 93, 98, 104, 110, 111–112, 130, 134, 155–156, 160, 163–167, 177, 182, 183
  data, *6*, *15*, *32*, *33*, *47*, *48*, *72*, *74*, *75*, *94*, *98*, *113*, *119*, *123*, *124*, *131*, *135*, *145*, *158*, 165, *181*
  definition of, 11
  multispecies, 14, *15*, 16–22, 24, *33*, 38, 97, *98*, *135*, 145, 163, 177–178, 181, 183, 184
trend, 166
Trichogramma, 29, *31*
Trichostrongylus tenuis, 112, 114, 124, 126
  aggregation of, 114
trophic interactions, 4, 10, 24

voles, 45–68, 162–163, 166, 167, 168, 178–179
Vulpes vulpes (*see* fox)

weasels (*see* mustelids)
weather effects, 34–35, 42 (*see also* climatic effects)
western pine beetle, 104–105, 181
winter moth, 150
winter temperature, 148–149
wolf, 80
wolverine, 80

Zeiraphera diniana (*see* larch budmoth)